# PROFESSIONALIZATION OF THE SENIOR CHINESE OFFICER CORPS

## *Trends and Implications*

JAMES C. MULVENON

T0159550

Prepared for the
Office of the Secretary of Defense

*National Defense Research Institute*

## RAND

# PREFACE

This report documents one component of a larger RAND effort to analyze key factors influencing China's emergent national security strategies, policies, and military capabilities, and their potential consequences for longer-term U.S. national security interests. Specifically, this report evaluates demographic and cohort changes in the officer corps of the Chinese People's Liberation Army (PLA), examining both cohort alignment and professionalizing trends in retirement, education, and functional specialization.

The report was produced under the aegis of a project entitled Chinese Global and Regional Strategy and U.S. Policy: Dynamics and Implications.

This research was sponsored by the Office of the Under Secretary of Defense for Policy. It was carried out under the auspices of the International Security and Defense Policy Center within RAND's National Defense Research Institute (NDRI), a federally funded research and development center sponsored by the Office of the Secretary of Defense, the Joint Staff, and the defense agencies. Supplemental funding was provided by the RAND Center for Asia-Pacific Policy.

# CONTENTS

# TABLES

Despite periodic instability, the expansion of military-to-military relations between the United States and China has constituted one of the more promising and important developments in Sino-U.S. relations in recent years. As these exchanges develop further, high-ranking U.S. military personnel and civilian policymakers will come into contact with an ever broader cross section of Chinese military personnel. To ensure maximum benefit from these contacts, U.S. officials will need to understand the changing composition and character of the highest levels of the Chinese officer corps. The attitudes and beliefs of these officers seem almost certain to exert growing influence on China's domestic and external behavior. This report examines long-term trends in Chinese military institutional development and assesses their impact upon the People's Liberation Army's (PLA's) future evolution.

Western scholars have long argued that the PLA officer corps since 1978 has become more professionalized. Specifically, they have asserted that the Chinese are implementing plans to make the army "younger in age, better educated, and professionally more competent."[1] A lack of hard data, however, has prevented rigorous analysis of demographic changes, especially the identification of important long-term professionalizing trends such as increased education levels or functional specialization within the officer corps. Using detailed biographic data sources, this report evaluates the scope and

---

[1]"PLA Officers Said 'Younger, Better Educated,'" Xinhua, 25 July 1997, in FBIS, 25 July 1997.

relative success of three professionalizing trends (education, functional specialization, and retirement norms) in the top 200 or so officers in the PLA. The data suggest the following conclusions:

- The base education level of the senior officer corps has increased between 1989 and 1994, and the new crop of officers has a higher rate of secondary, postsecondary, and advanced military education than its predecessors. More than 79 percent of the 1994 officer corps had received some form of advanced education, and more than 55 percent had received professional military education (PME). Furthermore, the curriculum of military education in China is now more technical and less political in focus, suggesting that professional norms are being inculcated among its attendees.

- The senior officer corps is increasingly specialized in functional expertise, with a clear differentiation among political, military, and technical officers. In 1994, 41 percent of PLA officers followed exclusively military careers (up from 33 percent in 1988), while the percentage of officers engaged only in political work rose from 27 percent in 1989 to 42 percent in 1994. In contrast, the percentage of officers whose career paths had involved both military and political work fell from 21 percent in 1989 to 8 percent in 1994. Overall, the percentage of officers engaged in specialized careers rose 26 percent, while those with cross-fertilized careers fell more than 68 percent. Finally, the distribution of specialized careers is highly correlated with age, as younger officers have followed more specialized careers than older officers.

- There is now a stable, functioning retirement system in the PLA based upon age, position, and rank. As a consequence, the average age of officers at every level of the PLA has declined. Between 1989 and 1994, the average age of a Central Military Commission (CMC) Member declined 13 years, General Department Director 9 years, and a Military Region Commander/Political Commissar 10 years. Significantly, the average ages in 1994 at each position are slightly less than 6 years below the prescribed maximum age for each respective position, in contrast to 1989, when the average ages were slightly less than 6 years above the prescribed maximum. Newly posted officers can therefore serve out the full term for a military region principal (3 to 5 years according to the "1994 Regulations for Active Duty Officers") without violating the strictures governing retirement age.

These trends represent positive developments for Chinese military professionalization since they improve the army's expertise, rationalize career patterns within the leadership, build corporateness, and reinforce the obedience of the military to its client, the Chinese Communist Party (CCP). They also confirm the technocratic transition occurring across the entire Chinese political and institutional system and provide further evidence of the transformation from revolutionary-era cadres to professional managers. However, it remains difficult to render a definitive judgment about the current level of professionalization in the PLA since the term includes many more dimensions than the available data permit (e.g., training). Nonetheless, the areas under examination show significant and lasting improvement and provide a solid base for the ongoing process of professionalization.

The report next evaluates cohort trends in the PLA. Cohort analysis identifies groups within the officer corps that, by virtue of common age or experience, share certain values or beliefs that can be compared with those of peers and nonpeers. Cohort analysis can be contrasted with factional analysis, which is concerned with identifying the distribution of political power within the military leadership, particularly with reference to the dynamics of civil-military relations. For the purposes of this report, our definition of a cohort centers on the general concept of group affiliation, which operates on a "we-they" distinction. Group affiliations within the officer corps include affiliations with military and nonmilitary groups. These associations contribute to the development of corporatism because they strengthen the organic unity and self-identification of the officer corps. Intramilitary affiliations can be both formal and informal. Formal in-service affiliations include relationships derived from the course of official duties, most importantly shared combat or staff experience. This is reflected in the conclusions relating to war experiences and the corps affiliations. Informal affiliations, on the other hand, include ties forged in quasimilitary functional associations and those developed during military education. These are examined in the context of shared educational experiences. The second major category of affiliations is nonmilitary groups, which includes preservice affiliations with a particular class or geographic area. To this end, we analyze the birth origins of the senior officer corps. Using both standard metrics (field-army affiliation) and newer frameworks (generational analysis), we reach the following preliminary conclusions:

- The field-army (FA) thesis continues to lose its explanatory power as the original revolutionary elders pass from the scene. Nonetheless, it still receives a great deal of attention among Chinese interlocutors who consistently stress its importance to foreign scholars. Affiliations with the 4th Field Army continue to dominate, although the strength of this group is diluted between the 4th FA and what is known as the "southern" 4th FA. If the 4th FA is divided along these lines, then the strongest single grouping is the 3rd FA, which makes up 28 percent of the total. The drop of the 3rd FA from 38 percent in 1989 to 28 percent in 1994 is surprising given the promotion in 1992 of one of its important veterans, Zhang Zhen, to the vice-chairmanship of the CMC with the portfolio for personnel decisions. Most striking, however, is the continued poor showing of Deng Xiaoping's 2nd FA, which dropped from 13 percent in 1989 to 9 percent in 1994.

- One alternative to the field-army thesis is the generational thesis, which divides the officer corps into various generational groups, each of which is assumed to have a set of common experiences and values. Nearly all of the current high-level military leadership are drawn from the same age cohort, whose key formative experiences included the Korean War, the Sino-Soviet alliance of the early 1950s, the 1962 Sino-Indian border war, the 1969 Sino-Soviet border clash, and the 1979 Sino-Vietnamese border war. Sixty-two percent of this group joined the PLA between 1945 and 1954, and 51 percent of them fought in the War of Liberation (1945–1949), the Korean War (1950–1953), or both. Only 32 percent, however, fought in the Korean War, making it inaccurate to refer to this generation as the "Korean War generation." It is further necessary to disaggregate this cohort into its constituent parts so that differences within the cohort (e.g., warfighters versus nonwarfighters) can be revealed. To this end, we examine a series of intergenerational variables, including birth origin, war experience, corps/group army affiliation, and professional military education.

- There are significant disparities in the distribution of geographic birth origin among the officer corps, but the best explanation seems to center on the geographic pattern of national unification during the late 1940s and early 1950s. For example, the regions that were liberated last (southwest, northwest, central-south) constitute by far the three lowest percentages (18.5 percent total), despite the fact that they contain large percentages of the overall population (49.8 percent). Those regions that were liberated earlier (north, northeast, east) constitute a much higher percentage of the officer corps (81.5 percent) than the overall population distribution (49.1 percent) would suggest. If this explanation is correct, however, we should expect that this distribution will even out over time.

- In terms of war experience, the data reveal a profound shift in experience from the Revolutionary War period (1927–1937) to the Civil War (1945–1949) and Korean War (1950–1953). In particular, the number of officers who had fought in the Anti-Japanese War (1938–1944) fell from 46 percent to only 2.5 percent, while the number of officers whose first combat experience was the Civil War or the Korean War rose from 30 percent to 51 percent. This change reflects the evolution of the PLA from a guerrilla army to a more modern fighting force. The legacy of these experiences for the current officer corps can be seen in many of the PLA's modernization and doctrine reforms in the 1980s, such as the change from "people's war" (*renmin zhanzheng*) to "limited war under high-tech conditions" (*jubu zhanzheng zai gaoji tiaojian*). Equally striking, however, is the increase in the number of officers who have no combat experience at all (from 21 percent to 46 percent), although this can be explained by age (the average officer in 1994 was 58 years old and would have been 14 at the time of the Korean War) and the rise of technocrats in the military leadership.

- In terms of PME attendance, less than 1 percent of the 1994 pool had attended the National Defense University (NDU) during Zhang Zhen's tenure. This either disproves the commonly held assumption that Zhang Zhen, in his current CMC position with the portfolio for promotions, has consciously promoted former NDU students that he met during his presidency, or exposes a serious gap in the data pool.

- The corps affiliations of officers in both pools were remarkably uniform, which suggests that no one corps or corps commander has exerted an overwhelming influence over promotion patterns in the PLA. Yet the evidence also suggests that the members of some corps are more successful than those of others. In the aggregate, 80 percent of the identifiable corps affiliations in the 1989 pool and 85 percent of the affiliations in the 1994 pool involve corps that survived the multitude of consolidations and deactivations following Liberation, the Korean War, and the demobilization and retirement of one million soldiers in 1985. While a minority of officers have been promoted despite the deactivation or consolidation of their "home corps," the vast majority of the sample rose through the ranks of corps that still exist in group-army form today. Overall, the corps/group-army variable has the greatest future potential of all the cohort variables examined in this study, particularly with regard to the connection between cohorts and networks. With better data, these linkages might be solidified, providing a powerful tool for analyzing internal Chinese military behavior.

Judging from these data, the PLA has clearly undergone a profound shift from the revolutionary generation to a new post-Liberation cohort. This cohort has more experience with modern warfare and is therefore inclined toward technological modernization and doctrinal evolution. These data also help clarify the personalistic dynamics in the PLA, which appear to be changing in character from the earlier faction-ridden era. Overall, the PLA could be said to be developing more professional-type networks (similar in some respects to the U.S. military) latticed around traditional personal ties as well as professional military education, field performance, and other avenues of professionalization. These networks have a contradictory effect upon the military: On the one hand, they strengthen the PLA by creating associational groups and providing additional sources of information for the promotion process; on the other, they weaken the PLA by creating new communication and power channels outside of the traditional chain of command. Some of these latter negative consequences, however, are mitigated by the fact that the professionalizing trends in the PLA assure that there is a rising meritocratic "floor," permitting networks to serve as a mechanism for differentiating among a cohort of largely professional and more competent officers.

On balance, therefore, the PLA is becoming more professionalized. The degree of PLA professionalization and the potential impact that the process of professionalization will have upon the Chinese military and political system is less clear since the process is also affected by the economic, political, and social trends in the nation as a whole. Yet it is indisputable that the PLA is now more professional than either its prereform or midreform counterparts, and this trend shows no signs of reversal.

These findings have important implications for future U.S.-Chinese military-to-military relations. Should contacts deepen to lower levels of the Chinese military system, U.S. officials and especially U.S. military officers will meet military officers more like themselves: professional, modern, well educated, and technically capable. Entrance requirements are much higher, as are the standards for promotion. As a result, the PLA officer corps should no longer be viewed in terms of its guerrilla origins, i.e., long on fervor but short on applicable skills. Instead, we should view the senior officer corps in the same way as we viewed the Soviet military, i.e., as a competent military to be respected—although this is not to say that the Chinese military is currently as capable as the Soviet Red Army was at its height or that its intentions are even remotely similar. The reform of the PLA has helped lay the groundwork for a military leadership capable of waging 21st-century warfare, even if its equipment still lags well behind advanced global levels. New equipment can be acquired, however, while professional officers capable of maximizing the value of that equipment must be slowly and patiently educated and trained in professional military education and field environment. It is these officers who will determine the quality and character of the Chinese military of the future.

# ACKNOWLEDGMENTS

This report benefited greatly from the environment in which it was created. In April 1995, Michael Swaine hired me as a Resident Consultant at RAND. The two years since have been the most stimulating and productive of my career, due in no small part to his professional guidance and personal friendship. I am indebted to him for the opportunity to study the Chinese military full-time and publish this report under the auspices of RAND. I am also very grateful to Michael's former research assistant, Kirsten Speidel, who warmly welcomed me to RAND and patiently answered my endless questions.

Other RAND colleagues cheerfully reinforced my belief that collegiality and professional excellence are not mutually exclusive. In particular, Jonathan Pollack, LTC William O'Malley (Ret.), Chung Min Lee, Rachel Swanger, Ashley Tellis, Adam Stulberg, and the "lunch group" (Ed Gonzalez, Joe Kechichian, Tom Szayna, David Ronfeldt) provided useful comments, suggestions, and encouragement, as well as their hearty friendship. I would like to extend my warm appreciation to Barbara Wagner, whom I burdened with thousands of annoying requests.

Special acknowledgment must be given to Ellis Melvin of Tamaroa, Illinois, whose tireless dedication to monitoring the rank-and-file of the PLA elevates him to the level of national security asset. I would also like to thank Jonathan Pollack and Paul H. B. Godwin for their detailed reviews of the final drafts of this report.

Finally, I would like to thank Mary Hampton for her enduring patience.

# ACRONYMS

| | |
|---|---|
| AMS | Academy of Military Sciences |
| CCP | Chinese Communist Party |
| CMC | Central Military Commission |
| COSTIND | Commission on Science, Technology, and Industry for National Defense |
| FA | Field Army |
| GLD | General Logistics Department |
| GPD | General Political Department |
| GSD | General Staff Department |
| MD | Military District |
| MR | Military Region |
| NDU | National Defense University |
| PAP | People's Armed Police |
| PLAAF | People's Liberation Army Air Force |
| PLA | People's Liberation Army |
| PLAN | People's Liberation Army Navy |
| PME | Professional Military Education |
| PRC | People's Republic of China |

# Chapter One

# INTRODUCTION

In the post–Deng Xiaoping era, the People's Liberation Army (PLA) will constitute one of the most important actors in Chinese politics. Many analysts believe that the PLA will play the role of "kingmaker" in the post-Deng succession, either actively lobbying for a particular aspirant or wielding veto power over potential challengers.[1] The changing makeup of the Chinese officer corps becomes critical in this context since the attitudes and beliefs of the PLA may significantly affect both China's domestic and external behavior. Various Western analyses argue that the PLA officer corps since 1978 has become more "professionalized."[2] They assert to varying degrees that the Chinese are steadily implementing a plan to make the army "younger in age, better educated, and professionally more competent."[3] A lack of hard data, however, previously prevented more rig-

---

[1]For example, see Michael Swaine, *The Military and Political Succession in China: Leadership, Institutions, Beliefs*, Santa Monica, CA: RAND, R-4254-AF, 1992, pp. 188–190.

[2]A few examples of this include Ellis Joffe, *The Chinese Army After Mao*, Cambridge, MA: Harvard University Press, 1987; Harlan Jencks, *From Muskets to Missiles: Professionalism in the Chinese Army, 1945–1981*, Boulder, CO: Westview Press, 1982; Monte Bullard, *China's Military-Political Evolution: The Party and the Military in the PRC, 1960–84*, Boulder, CO: Westview Press, 1984; Paul H. B. Godwin, *The Chinese Communist Armed Forces*, Maxwell Air Force Base, Alabama: Air University Press, 1988; Cheng Hsiao-shih, *Party-Military Relations in the PRC and Taiwan: Paradoxes of Control*, Boulder, CO: Westview Press, 1990; and June Teufel Dreyer, "The New Officer Corps: Implications for the Future," *China Quarterly*, No. 146, June 1996, pp. 315–335.

[3]"PLA Officers Said 'Younger, Better Educated,'" Xinhua, 25 July 1997, in FBIS, 25 July 1997. See Wang An, *Jundui zhengguihua jianshe [The Construction of Military "Regularization"]*, Beijing: Guofang daxue chubanshe, 1996. For a list of the official documents in which the policies on professionalization are verified, see Dong Lisheng,

---

orous analysis of demographic and cohort changes in the officer corps, especially the identification of important long-term professionalizing trends such as increased education level, functional specialization, and adherence to retirement norms.[4]

Li Cheng and Lynn White, in their 1993 *Asian Survey* article "The Army in Succession to Deng Xiaoping," sought to evaluate China's military elite transformation by analyzing newly available biographical data on the Chinese officer corps.[5] Although the study is path breaking in its use of new sources and statistical analysis, it has some serious methodological weaknesses. First, its two data sets are really not comparable. Specifically, the 1992 data set (made up of the 46 military members of the Fourteenth Central Committee) is in many ways a subset of the 1989 group (the 224 military entries from the 1989 edition of *Who's Who In China*),[6] thus making rigorous comparative analysis untenable. Second, Li and White do not place their results or analysis in any theoretical context. In particular, they do not review any of the literature on military professionalism or elite transformation.

The goal of this report is to enhance our understanding of Chinese military professionalization by correcting these two limitations. This report replaces Li and White's 1992 data set with military elite data found in the 1994 edition of *Who's Who in China*. The similar selection criterion of the 1989 and 1994 pools provides a homo-

guest editor, "The Cadre Management System of the Chinese People's Liberation Army (I) [*Zhongguo renmin jiefangjun ganbu zhidu gaiyao*]," *Chinese Law and Government*, Vol. 28, No. 4, July–August 1995, p. 50; and "Active Service Regulations Governing Active Duty Officers of the People's Liberation Army," Xinhua Domestic Service, 13 May 1994, in Federal Broadcast Information Service (FBIS), 17 May 1994, pp. 35–40.

[4]The exception is Swaine (1992), which contains an extensive analysis of high-ranking military elites.

[5]The full citation for this study is: Li Cheng and Lynn White, "The Army in Succession to Deng Xiaoping: Familiar Fealties and Technocratic Trends," *Asian Survey*, August 1993, pp. 757–786.

[6]*Who's Who in China* is published by the Foreign Language Press in Beijing and includes the top 2,000 or so military/political leaders in China. There have been only two editions of this source, one in 1989 and the most recent in 1994. See Liao Gailong and Fan Yuan, eds., *Who's Who in China: Current Leaders*, Vol. 3, Beijing: Foreign Language Press, 1989; and Liao Gailong and Fan Yuan, ed., *Who's Who in China: Current Leaders*, Vol. 4, Beijing: Foreign Language Press, 1994.

geneous pool of numbers for empirical evaluation.[7]  Together, the two data sets represent a hitherto unavailable level of disaggregation in military elite information, permitting differentiation between age cohorts of varying backgrounds, education level, and experience, as well as war experience and corps/field-army affiliation.

However, this report does not purport to build a theory of military professionalism in China.  The previous paucity of reliable data on the PLA has created a literature already replete with untested descriptive theories built upon Western studies of military professionalism and largely anecdotal evidence from limited primary and secondary sources.  In an effort to consolidate and strengthen the existing body of research on the Chinese military, this report proposes to use the newly available data to answer some basic questions about the transformation of the officer corps.  The analysis of these issues is divided into two chapters:  demographics and cohort analysis.  In the demographics chapter, we will attempt to answer the following sets of questions:

1. Is the senior officer corps of the PLA better educated than its predecessors, and, if so, by how much?  What is the content of that education?

2. Has the senior officer corps become more functionally special-ized, and, if so, how are those specializations differentiated?  What role does political work play in this division of labor?

3. Is there a functioning, age-based retirement norm in the PLA, and if so, what are its characteristics and parameters?  How widespread is the norm, and what explains the exceptions to the norm?

In the cohort analysis chapter, we will attempt to answer the follow-ing sets of questions:

1. How useful are our current models of cohort analysis in the PLA, in particular the traditional field-army thesis?  Are there any viable alternatives?  Are "generations" a useful category of analysis?

---

[7]These similarities will be outlined in the "Data and Methods" section.

2. What is the dominant "generation" of the current members of the military leadership, what were their formative experiences, and what are their shared values and beliefs?

3. What are the geographical origins of the current high-ranking officer corps, and what might explain patterns within the data?

4. What do the available data tell us about the correlation between professional military education and promotion?

5. What is the influence of corps/group-army affiliation on career patterns of the current military leadership?

This report is divided into four chapters. The remainder of Chapter One contains two short sections. The first is a discussion of data and methodology. The second briefly examines the theoretical assumptions and axioms of military professionalism. Chapters Two and Three contain the empirical evaluation of our demographic and cohort data, respectively. Chapter Four assesses the overall professionalism of the PLA in the light of our data; briefly explores the effect of professionalizing trends upon the PLA's relationship to the Party, the military's operational effectiveness, and the future of Sino-U.S. military relations; and then outlines avenues for future research.

## DATA AND METHODS

The two primary data sources for this report are the military entries from the 1989 (224 officers) and 1994 (179 officers) editions of *Who's Who in China*, published by the Foreign Language Press. Both data pools include all officers at or above the level of deputy chief of staff or political department deputy director in the seven military regions (Beijing, Shenyang, Jinan, Nanjing, Guangzhou, Chengdu, Lanzhou) and the three General Departments (General Staff Department, General Political Department, and General Logistics Department). Specifically, this includes all commanders, vice-commanders, political commissars, vice-political commissars, heads and vice-heads of staff, and heads and vice-heads of political departments at the military region level or above, as well as all directors, deputy directors, chiefs of staff, and deputy chiefs of staff of the three General Departments. Neither list is a statistical sample, and both may be analyzed as complete sets, although 53 percent of the military leadership in 1994 was also present in the 1989 pool. Nevertheless, this level of disaggregation allows for cross-sectional analysis of trends in the officer corps, especially the demographic and cohort-related differences between younger and older officers, including age, birthplace, education level, functional specialization, war experience, and field army/corps affiliation.

Regrettably, the nature of the data also creates some limitations on this research. First, the choice of variables is, by necessity, almost wholly data driven. Second, the report uses only open-source data, leaving gaps that might be correctable at a classified level. Third, the gaps in the data restrict the statistical operations that may be performed. Ideally, we would like to have employed sophisticated statistical tools in this research, regressing demographic variables on promotion to determine whether professional officers were being promoted over nonprofessionals. B. Mitchell Peck's work on the U.S. military is an important and useful example of this type of analysis.[8] Unfortunately, our data only describe the officers who were promoted and not those whose careers ended, leaving us with a self-selected group. Additionally, our data only describe officers at the military region and above, which is near the top of the military hierarchy. At that level, most of the weeding out in promotions and dismissals has already taken place.

A second inviting strategy would be to rank the 224 officers from the 1988 pool on an ordinal scale from 1 to 224 based upon their position and then regress their demographic characteristics (education level, etc.) against their status in the hierarchy. Li and Bachman were able to use this type of gamma-coefficient analysis on a single cohort of 247 Chinese mayors because they could rank cities by population size.[9] In the PLA, this type of ranking is extremely difficult, given the unavoidable element of subjectivity in assigning value to position. For example, it is very difficult to determine whether one deputy commander of a military region is "higher" than another or if the deputy commander of a military region is more or less important than the chief of staff of a general department. The mountain of as-

---

[8]B. Mitchell Peck, "Assessing the Career Mobility of U.S. Army Officers: 1950–1974," *Armed Forces and Society,* Vol. 20, No. 2, Winter 1994, pp. 217–237 (especially p. 219).

[9]See Li and Bachman, "Localism, Elitism, and Immobilism," pp. 80–81.

sumptions necessary to bolster this type of framework would create more problems than it would solve.

These types of methodological constraints are symptomatic of the much larger recurring problems in comparative studies of China. While the last five years have witnessed an unprecedented outpouring of new statistical and primary sources, there are still important gaps in the record. But different scholars have developed innovative research strategies. The work of Melanie Manion on cadre retirement in China and Andrew Wedeman's research on cadre corruption stand out in this regard.[10] Despite the limitations of the available data, however, the primary sources found in this report represent a significant improvement over previously available information on the Chinese military leadership and thus warrant close analysis.

## THEORETICAL PERSPECTIVES ON MILITARY PROFESSIONALISM

In order to analyze military professionalization in the Chinese case, it is necessary to define general terms and concepts. The classic study of military professionalism is Samuel Huntington's *The Soldier and the State*.[11] In Huntington's formulation, the modern warrior, evolved from the aristocratic dilettante, is committed to three distinguishing characteristics of the military profession as a special type of vocation: expertise, responsibility, and corporateness.[12] Expertise is defined as the functional knowledge gained through education and training that separates the professional from the layman.[13] Responsibility is the duty to selflessly apply these skills on behalf of the state.[14] Corporateness is the "sense of organic unity and consciousness" on the part of military officers that they are a

---

[10]Andrew Wedeman, "Bamboo Walls and Brick Ramparts," unpublished dissertation, University of California Los Angeles, 1995; Melanie Manion, *Retirement of Revolutionaries in China: Public Policies, Social Norms, Private Interests*, Princeton: Princeton University Press, 1993.

[11]Samuel Huntington, *The Soldier and the State: The Theory and Politics of Civil-Military Relations*, New Haven, CT: Yale University Press, 1957.

[12]Ibid., p. 8.

[13]Ibid., p. 8.

[14]Ibid., p. 9.

group apart from laymen, and is achieved through group discipline and training.[15]

Together, these three concepts form the core characteristics of a professional officer corps and serve as the theoretical foundation of our propositions on the professionalization of the Chinese military.[16] Each concept corresponds to a set of the questions raised in the introduction and will be examined in light of the available data, which are admittedly imperfect measures. The educational component of PLA expertise will be ascertained through educational levels and functional specialization within the senior officer corps, in particular the depth and content of professional military education (PME). Responsibility, or political obedience to the state (in this case the Chinese Communist Party [CCP]), will be addressed by evaluating age-based retirement norms and the political component of functional specialization. Finally, corporateness will be examined in both the demographic and cohort contexts. The chapter on demographics will assess the influence of standardized education, common school affiliation, and functional subprofessions upon the development of a more professionally-oriented corporatism within the PLA. In the chapter on cohort analysis, the corporate effect of shared service and internal military politics will be examined, especially the extent to which the affiliations of the PLA reinforce or undermine the professionalizing demographic trends.

---

[15]Ibid., p. 10.

[16]To those who would question the usefulness of Huntington in a Chinese context, Paul Godwin offers the following excellent synthesis: "In the profession of arms, the specialized and theoretical knowledge of the profession is universal, the ethical rules and doctrine tend to be particularistic, and the sense of corporateness, although universal, may originate in particularistic sets of doctrines and ethics." This modification allows the PLA to professionalize itself without rejecting the socio-cultural milieu in which it operates. It also permits redefinition of the term "professionalization" to include not only the dichotomy of "professional" versus "political," but also alternative (and in this case broader) definitions than simply the pure Huntingtonian model. Indeed, Godwin concludes that to deny the Chinese military is professionalizing simply because the process does not resemble its counterpart in the West is "an ethnocratic [sic] evaluation without sound analytical bases" which "distorts our analysis of Chinese politics and the role of the armed forces in such politics." See Paul H. B. Godwin, "Professionalism and Politics in the Chinese Armed Forces: A Reconceptualization," in Dale Herspring and Ivan Volyges, eds., *Civil-Military Relations in Communist Systems*, (Boulder, CO: Westview Press, 1978), pp. 219–240.

Virtually all previous studies of the Chinese military agree that the post-Mao PLA officer corps is becoming more professionalized in Huntington's sense of the term.[17] Disagreement centers on the pace and scope of professionalization, and the extent to which political control of the military by the CCP impedes or enhances it. At the center of the argument is Ellis Joffe.[18] In *The Chinese Army After Mao*, Joffe argues that the reforms by the late 1980s had transformed the military into a professional army, with increasingly less interference from ideological elements inside or outside the army. Joffe's critics take issue with both points, arguing that while the professionalization of the PLA has achieved some success, political, ideological, and/or logistical considerations have seriously impeded its progress.[19] In recent years, Joffe himself seems to have qualified his earlier conclusions, pointing to the rampant corruption and commercial activities of the PLA as evidence of retarded professionalization.[20]

The paucity of hard data on the PLA, however, has forced all participants in this debate to rely on limited primary and secondary source material, such as official Chinese media, Hong Kong/Taiwan-based journalism, and some modest direct observation of trends in Chinese military-institutional development. With the new data, however, it is now possible to evaluate empirically (although admittedly in a limited sense) some of the debate's key questions as well as the claims of the Chinese themselves. Given that the 1989 and 1994

---

[17]See footnote 2. I have intentionally separated the question of post-Mao professionalization from the debate in the literature about the nature of the PLA during the Maoist era. Works by Zhu Fang, Thomas Bickford, and others make compelling arguments about PLA professionalism during those earlier periods that promise to permanently alter our view of the army under Mao. This report does not intend to address these issues. Instead, it looks only at professionalization since 1978, in particular the claims of Western scholars like Joffe and the Chinese themselves, and seeks only to evaluate their validity.

[18]See Ellis Joffe, *Party and Army: Professionalism and Political Control in the Chinese Officer Corps, 1949–1964*, and *The Chinese Army After Mao*.

[19]See Paul H. B. Godwin, "Professionalism and Politics in the Chinese Armed Forces: A Reconceptualization," pp. 219–220; and Jonathan Pollack's review of Joffe's book in the *American Political Science Review*, Vol. 84, No. 1, March 1990, pp. 339–340.

[20]Ellis Joffe, "The PLA and the Economy: The Effects of Involvement," IISS/CAPS Conference, "Chinese Economic Reform: The Impact on Security Policy," Hong Kong, 8–10 July 1994.

biographical information  primarily involves demographics, it seems best suited to answer questions about the changing composition of the officer corps, particularly with regard to education level, functional specialization, and retirement norms.  In essence, these variables reveal trends in the changing makeup of the officer corps, which are in turn indicators of the military's institutional interests and the more macro-level changes in the society from which the officers are recruited.[21]  Such demographic analysis has yielded fascinating results on Chinese elite transformation, revealing the increasingly technocratic nature of civilian cadres and the extent to which technocratic affiliations have begun to replace traditional criteria for promotion.[22]  In particular, these studies have statistically documented the transition from revolutionary cadres to new professional managers.  When these changes are compared to those occurring in the PLA, one preliminary conclusion is that the differing demands on civil and military leaders are dividing the two elites, with one set specializing in the intricate tasks associated with leading and managing an increasingly complex civil society and the other developing the skills required for managing and directing an increasingly complex military system.[23]

---

[21] The relationship between elite change and societal change is discussed in Li Cheng and David Bachman, "Localism, Elitism, and Immobilism: Elite Formation and Social Change in Post-Mao China," *World Politics*, Vol. 42, No. 1, October 1989, pp. 64–94.

[22] A few examples of this literature are Andrew G. Walder, "Career Mobility and the Communist Political Order," *American Sociological Review*, Vol. 60, June 1995, pp. 309–328; Li Cheng, *The Rise of Technocracy: Elite Transformation and Ideological Change in Post-Mao China*, Dissertation, Department of Politics, Princeton University, 1992; Li Cheng and Lynn White, "Elite Transformation and Modern Change in Mainland China and Taiwan: Empirical Data and the Theory of Technocracy," *China Quarterly*, No. 121, March 1990; Hong Yung Lee, "Mainland China's Future Leaders: Third Echelon of Cadres," *Issues and Studies*, Vol. 24, No. 6, June 1988, pp. 36–57; Li Cheng and Lynn White, "The Thirteenth Central Committee of the Chinese Communist Party: From Mobilizers to Managers," *Asian Survey*, Vol. 28, No. 4, April 1988, pp. 757–786; Hong Yung Lee, "China's 12th Central Committee: Rehabilitated Cadres and Technocrats," *Asian Survey*, Vol. 23, June 1983, pp. 673–691; William deB. Mills, "Generational Change in China," *Problems of Communism*, Vol. 32, November–December 1983, pp. 16–35; and Monte Bullard, "People's Republic of China Elite Studies: A Review of the Literature," *Asian Survey*, Vol. 19, No. 8, August 1979, pp. 789–800.

[23] I would like to thank Paul Godwin for bringing this point to my attention.  This argument is also presented in Ellis Joffe, "Party-Army Relations in China:  Retrospect and Prospect," pp. 303–304.

At the same time, the technocratic changes in the PLA do not obviate the need for continued cohort analysis of the senior officer corps. Even more fully professionalized militaries operate under systems of promotion and advancement that rely on both objective criteria and personalistic considerations. But nonprofessional considerations do not necessarily undermine the professionalism of the PLA writ large. Indeed, a certain percentage of personnel decisionmaking must be based on subjective criteria in order to ensure the corporateness leg of Huntington's professional triangle. In the United States military, for instance, affiliations based upon shared professional military education, while vehicles for personalistic association, also provide high-ranking officers with additional information vital to the promotions process.[24] Thus, any discussion of professionalism in the Chinese military must take into account both technocratic and cohort-related changes.

---

[24]PME, professional military education, is a U.S. military term that refers to a range of military training and educational programs, offered at all levels of the forces. This term is used in this report because the terms "academy" or "school" have different meanings in the Chinese and American contexts, and thus would be confusing to the Western reader.

# Chapter Two

# DEMOGRAPHIC CHANGES

Peacetime education and training should be considered a matter of strategic importance.—Deng Xiaoping, 1975

The quality of a commander determines to a great extent the quality of the troops, whereas the standard of colleges and academies determines the quality of commanders. . . . The relationship between the academies and troops is like that between the head and the body of a dragon while performing a dragon dance. If the dragon head performs well, its long body will soar aloft and dance freely in the air. In the course of building a modern, regular and revolutionary army, the building of colleges and academies is the key link that determines the building of the army.—Xiao Ke[1]

## EDUCATION

Huntington and other theorists have long argued that professional expertise in any field can only be achieved through prolonged education and experience.[2] In the military context, this education is provided by institutions of higher learning, which teach, preserve, and develop the body of military knowledge. At each stage of an officer's career, he is required to continue his education, receiving an increasingly macro-level view of the military and civilian world, as dictated by the needs of his future postings. An extremely important

---

[1]Xiao Ke, "Guidelines for Building Colleges and Academies of Our Army," *Renmin Ribao*, 3 October 1983, in FBIS, 5 October 1983, p. K8.

[2]This discussion is drawn largely from Huntington, *The Soldier and the State*, pp. 8–14.

aspect of this professional knowledge is its rigorous standardization because it creates a lingua franca for communication between members of the profession. This common language also serves to reinforce corporatism by instilling a bond of common understanding among graduates and transmitting a set of military norms for group behavior.

In the post-Mao military reforms, educational reform occupies a central place. In his 1977 speech "The Army Should Attach Strategic Importance to Education and Training," Deng Xiaoping bemoaned the fact that the PME of the PLA was in a shambles, leaving the Chinese officer corps "deficient in the ability to direct modern warfare."[3] His remedy called for not only the reopening of schools closed during the Cultural Revolution but an enlargement of the entire military education system.[4] In particular, Deng stated that military education and training must be raised to a "strategic position" and called for comprehensive training from the platoon level to the highest echelons of the military leadership:

> ... we must consider educating all officers, from platoon leaders up, in officers' training schools. Platoon or company officers should be graduates of junior infantry schools. ... Battalion and regimental cadres should go through intermediate officers' training schools. ... Likewise, leading cadres at the army or divisional level should be appointed only after they have attended senior officers' training schools.[5]

After 1978, Deng Xiaoping's proposed education reforms were quickly translated into identifiable policies. At the Third Plenary Session of the Eleventh Central Committee in 1978, the Central Military Commission (CMC) codified the relationship between promotion and education level, declaring that "those who have not received training in military academies cannot be promoted."[6] As a result, the

---

[3]Deng Xiaoping, *Selected Works of Deng Xiaoping*, Vol. 1, Beijing: Foreign Language Press, 1982, p. 75.

[4]Ibid., pp. 75–79.

[5]Ibid., p. 274. See also Jencks, *From Muskets to Missiles*, p. 57.

[6]See "Active Service Regulations Governing Active Duty Officers of the People's Liberation Army," p. 35; and Hsiao Chung, "Former Commander of Chengdu Military

officer training program of the PLA was reestablished, and new officers were required to attend some form of PME. Since many active-duty officers were unable to meet these requirements, supplemental courses called "spare-time universities" or "cadre cultural schools" were used to create a minimum standard.[7] Those who were still unable to meet these new standards were relieved, demoted, or forcibly retired.

This sea change encountered significant opposition from many active-duty officers promoted during the Cultural Revolution on the basis of their political qualifications. To overcome this resistance, the General Political Department issued a decision in 1983 stating that henceforth an officer's educational record would be as important as experience and performance in determining placement and promotion.[8] The same report stipulated that 70 percent of officers of platoon rank and above, as well as all commanders of naval vessels and pilots, would have to undergo PME.[9] Adding more specificity, officers had to be graduates of senior middle schools and had to pass an entrance examination given by the Ministry of Education.[10]

The resistance among members of the officer corps to these changes was indirectly aided by the chaotic condition of the nation's military educational system. The Cultural Revolution had destroyed much of China's professional military education system. Of the 140 institutions that were open before 1965, only 40 were intact in 1977.[11] In

---

Is Sent to National Defense University for Further Study," *Kuang Chiao Ching*, No. 233, 16 February 1992, pp. 14–17, in FBIS, 3 March 1992, pp. 34–35.

[7]Lonnie Henley, "Officer Education in the Chinese PLA," *Problems of Communism*, May–June 1987, pp. 55–71.

[8]Zhu Ling, "China's Army is Gearing Itself for Modern Warfare," *China Daily*, 11 June 1983, in FBIS, 13 June 1983, p. K30; "PLA Political Unit Makes Decision on Education," Xinhua, 4 May 1983, in FBIS 10 May 1983, pp. K8–9.

[9]Previously, Chinese officers had been promoted mainly from the troops. However, under the 1984 military service law, military academies have become the main source of officer material. See "Military Urged to Support Academy Reform," Xinhua, 2 June 1986, in FBIS, 3 June 1986, pp. K25–26.

[10]See Dong Lisheng, "The Cadre Management System of the Chinese People's Liberation Army (I)," p. 39. In 1986, the People's Liberation Army Air Force (PLAAF) began to enroll only pilots who passed national entrance exams. See "Half of All Pilots Have College Diplomas," Xinhua, 25 May 1995, in FBIS, 25 May 1995, p. 42.

[11]Liu Huinian and Zhang Chunting, "To Run the Army Well, It Is First Necessary to Run the Military Colleges Well—Xiao Ke on Building of Military Colleges," *Liaowang*,

the early 1980s, Xiao Ke described the fate of the PLA Military Academy during the ten years of chaos:

> Its teaching staff was disbanded. Its teaching materials and files were bundled into sixteen trucks and were reduced to ashes. Books in its library and its teaching equipment were practically all demolished.[12]

Thus, China's military schools had be rebuilt and many new ones established if the leadership hoped to accommodate the new legions of students.[13] The sheer enormity of the task caused at least one scholar, writing in 1981, to express open skepticism about Deng Xiaoping's military education goals: "It remains to be seen whether these ambitious standards can be met in 1985, or later."[14]

Within several years, however, the percentage of officers with advanced education began to increase. According to a 1987 *People's Daily* article, the proportion of PME graduates or college-educated officers increased from 1 percent in 1982 to 58 percent by 1987 at the army level, 2 percent to 66 percent at the division level, and 2 percent to 41 percent at the regiment level.[15] Table 2.1 shows that by 1994 college-educated cadres dominated even the highest levels of the military: More than three-quarters (79 percent) of officers in the samples had undergone some form of advanced education, up from 75 percent in 1989.

While this distribution of advanced education in the Chinese officer corps is lower than comparable groups in the Soviet (90 percent) or U.S. (97 percent) military, the profile is significantly higher than that

---

No. 7, 20 July 1983, in (Joint Publications Research Service) JPRS-CPS-84-273, No. 454, 8 September 1983, p. 74.

[12]Ibid.

[13]According to a Xinhua report, more than 32 percent of the army's total construction budget between 1978 and 1986 was devoted to building and renovating existing military schools. See Wang An and Yang Minqing, "PLA Modernizes Military Education System," Xinhua Domestic Service, 14 September 1986, in FBIS, 17 September 1986, p. K1.

[14]Jencks, *From Muskets to Missiles*, p. 227.

[15]*Renmin Ribao*, Overseas Edition, 6 July 1987, p. 1.

Table 2.1

Education Level of China's Military Leaders

|  | 1989 | | 1994 | |
|---|---|---|---|---|
|  | No. | Percent | No. | Percent |
| Type of Education |  |  |  |  |
| Professional military education (PME) | 116 | 51 | 98 | 55 |
| Military technical school | 37 | 17 | 33 | 18 |
| Nonmilitary college | 15 | 7 | 12 | 6 |
| All schools | 168 | 75 | 143 | 79 |
| None | 56 | 25 | 36 | 21 |
| Total | 224 | 100 | 179 | 100 |
| Location of Institution |  |  |  |  |
| Domestic | 155 | 92 | 134 | 94 |
| Foreign |  |  |  |  |
| USSR | 12 | 7 | 7 | 5 |
| USA | 1 | 1 | 0 | 0 |
| Unknown |  |  | 2 | 1 |
| Total | 168 | 100 | 143 | 100 |

SOURCE: Mulvenon PLA database.

of the prereform PLA.[16]  For example, a 1985 Xinhua broadcast asserted that the number of Navy commanders who were graduates of military colleges and universities was nine times that in 1965.[17] As recently as 1982, the number of officers at the military district level or above with any advanced education was just 4 percent.[18]

These rates of advanced education among the officer corps correlate with a prior stage of civilian cadre professionalization (*zhiyehua*), which began earlier than the military transformation.  For example, the 1994 rate of military education (79 percent) is comparable to 1984 levels among civilian cadres.  According to Li and Bachman, the percentage of college-educated cadres among various People's Re-

---

[16]The data on the U.S. and Soviet officer corps can be found in Gwendolyn Stevens, Fred Rosa, Jr., and Sheldon Gardner, "Military Academies as Instruments of Value Change," *Armed Forces and Society*, Vol. 20, No. 3, Spring 1994, pp. 473–484.

[17]See "Education Termed 'Vital' to Modernization," Xinhua, 26 July 1985, in JPRS-CPS-85-085, 22 August 1985, p. 76.

[18]*Renmin Ribao*, Overseas Edition, 6 July 1987, p. 1.

public of China (PRC) leadership levels in 1984 was as follows: Polit-buro, 67 percent; CCP Secretariat, 83 percent; CCP Central Committee, 73 percent; Ministry, 71 percent; Province, 62 percent; and Municipality/Prefecture/County, 78 percent.[19]  The percentage of college-educated cadres at each of these levels is almost 100 percent now and foreshadows one possible future for the officer corps.[20]

The data in Table 2.1 also show that 55 percent of the 1994 officers (up from 51 percent in 1989) had received PME from institutions like the PLA Military Academy or the National Defense University (NDU).[21] This trend correlates with the U.S. officer corps in the early 1950s, when 51 percent of all high-ranking officers were PME trained.[22]   It falls short, however, of the goal stipulated in the February 1984 General Political Department circular "Seven Year Plan on the Building of Leadership Groups and the Four Transfor-mations of the Cadre Contingent," which aimed for a 100 percent PME graduation rate among members of leading bodies of large units (*dadanwei*) by 1990.[23] Admittedly, the goal of universal PME attendance was probably unrealistic given logistical and political considerations, but this should not detract from the remarkable progress achieved thus far.

The data also reveal that more than 18 percent of the 1994 officer corps attended technical schools, such as artillery and engineering schools, and 6 percent attended a nonmilitary college.  This latter

---

[19]Li and Bachman, "Localism, Elitism, and Immobilism," p. 74.

[20]See Li Cheng, *The Rise of Technocracy: Elite Transformation and Ideological Change in Post-Mao China.*

[21]Within certain modernization-oriented service branches, the percentages are even higher.  In 1990, for example, it was reported that 73 percent of warship captains and their immediate deputies in the PLA Navy had received training in military academies. See Li Cheng and Lynn White, "The Army in the Succession to Deng Xiaoping," p. 780. Also, it must be pointed out that the PLA Military Academy ceased to exist after 1985.

[22]Morris Janowitz, *The Professional Soldier: A Social and Political Portrait*, Glencoe, IL: The Free Press, 1971, pp. 127, 140; and David Segal, "Selective Promotion in Officer Cohorts," *Sociological Quarterly* 8 (1967), pp. 199–206.  By 1964, the percentage of academy-trained officers in the U.S. military had risen to 75 percent.

[23]*Dadanwei*, or large units, refers to the service branches, large military regions, three general departments, and other units of equivalent level, which closely corresponds with our data set. For discussion of this General Political Department (GPD) circular, see Dong Lisheng, "The Cadre Management System of the Chinese People's Liberation Army (I)," p. 56.

percentage is certain to rise in the coming years, as the PLA has made serious efforts to recruit college graduates from civilian universities in the hopes of eventually raising the pre-PME education level in the officer corps from middle school to high school or even college.[24] In particular, the State Council and the CMC in March 1983 approved a circular entitled "Report on Assigning Some University Graduates to the Army to be Trained as Military and Political Commanding Cadres," which stipulated that a certain number of graduates from civilian universities would be assigned to work in the army.[25] They were to undergo short-term training at junior command schools before being assigned to platoon-level units as leaders.  Official documents assert that this group was to be "fostered" in the hopes that they would rise to high-level military and political posts.[26]

The evidence also suggests that the transformation of educational levels in the officers corps was not simply a function of one or two mass retirements of uneducated officers.  If we break down the distribution of military education by age groups, as seen in Table 2.2, it appears that younger officers continue to be better educated than their older counterparts.  In both 1989 and 1994, a greater percentage of officers between the ages of 45 and 57 had attended a command school, military technical school, or nonmilitary college than had those aged 58 to 63 or 64 to 75.  At the same time, the percentage of officers in the youngest age cohort who had attended one of the three institutions increased from 78 percent to 94 percent between 1989 and 1994.  More important, the percentage of officers with no higher education whatsoever declined precipitously for all but one age group between 1989 and 1994, falling 20 percent for officers aged 45 to 57 and 64 to 75.  The holdovers from the old system explain the anomalous increase in the rate of noneducated officers aged 58 to 63

---

[24]"PLA Attracts Graduates With Advanced Degrees," Xinhua, 22 July 1994, in FBIS, 22 July 1994, p. 20. For a discussion of the absorption of civilian cadres and graduates of civilian universities and secondary specialized schools, see Dong Lisheng, "The Cadre Management System of the Chinese People's Liberation Army (I)," pp. 40–41.

[25]Ibid., p. 41.

[26]Ibid.

## Table 2.2

## Education Level of the PLA Officer Corps, by Age Group

| | 1989 | | | | 1994 | | | |
|---|---|---|---|---|---|---|---|---|
| | 45–57 | 58–63 | 64–75 | All | 45–57 | 58–63 | 64–75 | All |
| Educational Attainment | No. (%) | No. (%) | No. (%) | No. (%) | No. (%) | No. (%) | No. (%) | No. (%) |
| Military school (see below for disaggregation) | 42 (71) | 81 (72) | 30 (56) | 153 (68) | 27 (85) | 58 (70) | 46 (72) | 131 (73) |
| Nonmilitary college | 4 (7) | 9 (8) | 2 (4) | 15 (7) | 3 (9) | 3 (4) | 5 (8) | 11 (6) |
| All schools | 46 (78) | 90 (80) | 32 (60) | 168 (75) | 30 (94) | 61 (74) | 51 (80) | 142 (79) |
| None | 13 (22) | 22 (20) | 21 (40) | 56 (25) | 2 (6) | 22 (26) | 13 (20) | 37 (21) |
| Total | 59 (100) | 112 (100) | 53 (100) | 224 (100) | 32 (100) | 83 (100) | 64 (100) | 179 (100) |
| Military schools | NA[a] | NA | NA | NA | (N=27) | (N=58) | (N=46) | (N=131) |
| Military academy | NA | NA | NA | NA | 19 (60) | 42 (51) | 37 (58) | 98 (55) |
| Military tech school | NA | NA | NA | NA | 8 (25) | 16 (19) | 9 (14) | 33 (18) |

SOURCE: Mulvenon PLA database.
[a]NA = Not Available.

from 1989 to 1994. The 26 percent of uneducated officers aged 58 to 63 in 1994 is almost exactly the same group as the 22 percent of uneducated officers aged 45 to 57 in 1989. Because of their experience and stature, these officers continued to be promoted through the system. The future, on the other hand, can be glimpsed in the change in educational levels of the youngest officers entering the ranks of the military leadership from 1989 to 1994, when the rate of noneducation plummeted from 22 percent to 6 percent. Taken together, these data strongly suggest that China's military education reforms are proceeding apace and offer no indication that the rates will not continue to rise as the older officers retire.

Equally important, however, is the nature and focus of PME. Only if the curriculum of the PLA officer corps has been substantively changed to reflect the new emphasis on professionalism can the higher rate of education among the military leadership be used as evidence of further professionalization. After the Cultural Revolution, the Chinese military educational system was divided into three tiers, reflecting increasingly higher stages of education. At the lowest level are the regional military schools, which provide a curriculum that combines an undergraduate college education with military training. The student body is drawn from two sources: senior high school graduates and promising enlisted soldiers. To enter the regional PME system, students are required to pass competitive national examinations. On the second tier, midlevel officers selected for promotion, usually captains or majors, are sent to a command college for a one-year course, after which they return to their units as battalion commanders.

At the highest level is the NDU. The NDU was formed in November 1985 by merging the PLA Military Academy, the PLA Political Academy, and the PLA Logistics Academy.[27] Its official mandate is to train commanders at and above the division level and staff officers at and above the military region level to "face the world and face

---

[27] Luo Tongsong, Wang Jin, and Gai Yumin, "Leaders Attend Founding of Defense University," Xinhua, 15 January 1986, in FBIS, 17 January 1986, pp. K5–K6; Wang Xian, "Analysis of the Chinese Communists' Establishment of a 'National Defense University,'" *Studies on Chinese Communism*, Vol. 20, No. 8, 15 August 1986, pp. 83–90, in JPRS-CPS-86-078, 9 October 1986, pp. 1–15; and Dai Xingmin and Gai Yumin, "China's Highest Military Institution—Visiting the National Defense University," *Ban Yue Tan*, No. 18, 25 September 1986, pp. 44–47, in FBIS, 15 October 1986, pp. K13–15.

the future."[28]  It was also specifically created to introduce "joint" (multiservice) education into senior PME, which at lower levels is service- and branch-based, in order to prepare senior officers for high-level command and staff positions where joint operations are now of central importance.[29]  In form and content, the NDU therefore resembles Western military institutions like the United States National Defense University or the General Staff Academy in the former Soviet Union, both of which perform educational roles as well as introduce civilian and military elites to the interagency nature of national security decisionmaking and policy implementation.[30]

The NDU offers two-year courses for officers chosen to command divisions, a one-year course for division and Group Army (GA) commanders who will soon be promoted to flag rank, and a three-month capstone course for both senior-level military and civilian personnel. Students are judged by their knowledge of military science as well as their mental, physical, and moral fitness, and those who receive outstanding ratings are recommended to the CMC and General Political Department (GPD) for promotion outside of "normal routes."[31] Between 1986 and 1994, the NDU trained more than 4000 students, of whom 147 have been promoted to deputy army leadership posts, 119 to army leadership posts, 51 to deputy leadership posts in military regions, and 13 to leadership posts in military regions.

Of particular interest is the NDU's capstone course, which is usually held in autumn and lasts approximately 100 days.  The course

---

[28]Gai Yumin and Xiong Zhengyan, "National Defense University, China's Highest-Level Military Academy, Founded in Beijing," Beijing Hong Kong Service, 18 December 1985, in FBIS 18 December 1985, p. K1; and Hsiao Chung, "Former Commander of Chengdu Military Is Sent to National Defense University for Further Study," p. 35.

[29]I would again like to thank Paul Godwin for bringing this point to my attention.

[30]In the early 1980s, representatives of the three general departments and the three military academies visited military academies in the United States, Britain, France, Germany, and Italy in order to study their organizational pattern, educational facilities, and curricula.

[31]This circumvention of "normal" promotion channels might have both positive and negative consequences for professionalism.  On the positive side, this type of fast track might increase elite mobility within the army, permitting promising young officers to bypass the constraints of seniority.  On the other hand, this policy might create an institutionalized avenue for patronage and factionalism, allowing senior officers to single our potential protégés.

focuses on policy issues, especially those related to national security. According to Dreyer, the average class size is 40, of whom 35 will be corps commanders and deputy regional commanders from both the PLA and the People's Armed Police (PAP).[32] All students are flag-rank officers who have served in their present capacity for several years. To maintain an equitable distribution of candidates, a balance between military regions and service headquarters (ground forces, navy, air force) is sought, with each region and headquarters choosing its own representative.   Students are encouraged to research military strategy and the international security environment in the hopes that some will become national-level military and political leaders.[33]

The curricula of these advanced command schools have also been drastically upgraded to meet the needs of a modernizing army.[34] In the early reform period, there were persistent criticisms of the military's education system, including the complaint that the "courses offered by the academies followed the content of [military academies in] the 1950s and 1960s."[35] In contrast, the new curricula seek to fulfill goals outlined in the 1994 "Regulations on PLA Active Duty Officers":

> [The PLA officer] must have the theoretical level [of knowledge], ability to understand policy, scientific knowledge, education level, professional knowledge, and organizing and command ability to carry out the assigned job.[36]

---

[32]Dreyer, "The New Officer Corps," p. 320.

[33] Xu Jingyue and Jing Shuzhan, "PLA Adopts System For Promoting High-Ranking Cadres," Xinhua Domestic Service, 24 June 1994, in FBIS, 28 June 1994, pp. 41–42.

[34]See Wang An, *Jundui zhengguihua jianshe [The Construction of Military "Regularization"]*; Hao Huaizhi, ed., *Dangdai Zhongguo jundui de junshi gongzuo [Contemporary Chinese Military Work]*, (Beijing: Zhongguo shehui kexue chubanshe, 1989, pp. 82–89, 342–347; and Dai Xingmin and Gai Yumin, "China's Highest Military Institution," p. K14.

[35]"Be Farsighted in Investing in Trained Persons," *Jiefangjun Bao*, 22 February 1983, p. 1.

[36]"Active Service Regulations Governing Active Duty Officers of the People's Liberation Army," p. 36.

In the beginning stages of reform, the curricula were narrowly focused, since the political battles between the reformers and more conservative elements had yet to be resolved. As a result, the NDU's first series of graduate student classes in 1986 focused only upon three disciplines: military thought, military campaign studies, and the history of warfare. As reforms progressed further, however, the high-level PME curricula became more specialized. For example, the Naval Command Academy in Nanjing in 1992 offered more than 53 subjects in 79 courses, including "military theory," "modern naval equipment," "naval command science," "military economics," "military psychology," and "electronic countermeasures by naval vessels."[37] Conversations with knowledgeable officials confirm that this trend is being replicated throughout the system.[38] This suggests that the curricula of China's highest military schools are becoming increasingly professional in orientation, and that the PLA's higher rates of education are enhancing the professional expertise of the officer corps.

At the same time, increased attention to professional subjects has not excluded political and ideological classes for the modern officer corps, though the number and focus of these courses have been drastically altered. In 1985, high-level military students at the NDU studied Marxist philosophy, Marxist political economics, and Party history.[39] By 1992, Naval Command Academy students were studying civil-military relations in foreign countries, and political study sessions focused more on the army's loyalty to the party than the abstract ideological principles of Marxism-Leninism-Mao Zedong thought.[40] On a substantive level, these are fundamental shifts in the ideological focus of PLA military education and reflect the ongoing transformation of the Chinese military from Mao's "armed body [for] carrying out the tasks of the revolution" to a professional

---

[37]Zhang Zenan, "Rear Admiral Li Dingwen, Director of the Naval Command Academy, Speaks of the Chinese Navy's Highest Institution of Higher Learning," *Jianchuan Zhihshi [Naval and Merchant Shipping]*, No. 8, 8 August 1992, pp. 2–3, in JPRS-CAR-93-001, 8 January 1993, pp. 51–54.

[38]James Mulvenon, personal interviews in Beijing, March 1997.

[39]Wang Xian, "National Defense University," p. 9.

[40]Zheng Zenan, "Naval Command Academy," p. 53.

military assigned to deal with high-tech warfare.[41]   Whereas the former required intense political indoctrination at all levels, the latter needs a strict division between military and political roles, using the political departments of units for propaganda and morale work rather than operational guidance.   The repoliticization campaigns following Tiananmen were an anomaly in this trend, but they were largely completed by January 1990.   After that date, professional themes once again balanced political themes in the military media and political study, and the normal training cycle was resumed.[42]

However, the reform of the military education system might also have unintended consequences for the development of pure Huntingtonian professionalism in the PLA.   In particular, a common school tie or class ring could make graduates a more cohesive pressure group and therefore more politically powerful.[43]   For the PLA, this is a potentially new source of military networks, although distinct in character from traditional factional wellsprings like the field-army system.[44]   Factions, of course, are unprofessional because their structure threatens the vertical authority of military hierarchy and the chain of command.   Networks, however, are an established component of professional militaries.   For example, in the United States military it is widely believed that "old boy networks" coalesce around common attendance at the U.S. Military Academy at West Point or the Naval Academy at Annapolis.[45]

---

[41]This quote can be found in Mao Zedong, *Selected Works of Mao Zedong*, Vol. 1, Beijing: Foreign Language Press, 1967, p. 106. The doctrines of "people's war under modern conditions" and later "limited wars under high-tech conditions," on the other hand, are credited to Deng Xiaoping. See Liu Yuming, "On Persistently Exploring People's War on High-Tech Terms," *Guofang [National Defense]*, 15 October 1994, No. 10, p. 8, in FBIS, 18 September 1995, pp. 19–20.

[42]For an excellent discussion of the ebb and flow in post-Tiananmen political work, see David Shambaugh, "The Soldier and the State in China: The Political Work System in the People's Liberation Army," *China Quarterly*, No. 127, September 1991, pp. 527–568.

[43]This argument has been made most forcefully by Bengt Abrahamsson, *Military Professionalization and Political Power*, Beverly Hills, CA: Sage Publications, 1972.

[44]See Tai Ming Cheung, "Back to the Front: Deng Seeks to De-Politicize the PLA," *FEER*, 29 October 1992, pp. 15–16; and Li Cheng and Lynn White, "The Army in the Succession to Deng Xiaoping," p. 781.   In fact, Li and White have argued (pp. 760–761) that "school ties" will be the field army paradigm of the 21st century PLA. See Ibid.

[45]For a survey and statistical evidence of this phenomenon, see B. Mitchell Peck, "Assessing the Career Mobility of U.S. Army Officers: 1950–1974," p. 219; and

On the other hand, the term "old boy network" unfairly implies that this structural development is a wholly negative phenomenon. In fact, some influential scholars argue these PME-based networks can also be beneficial to the functioning of a professional military since graduates "bring a special quality to the service that allows the military to maintain its distinctive military character."[46] Janowitz, for example, argues that attendance at military schools is the "source of the pervasive 'like-mindedness' about military honor and for the sense of fraternity which prevails among military men."[47] Likewise, Segal argues that systematically excluding officers without academy backgrounds is "functional for the maintenance of the structural autonomy of the military system" in the face of increasing civilian penetration of the military.[48] Finally, attendance at a military academy might fuel what David Moore and B. Thomas Trout term the "visibility theory of promotion," providing officers with opportunities to demonstrate their abilities and hence their suitability promotion.[49] Specifically, networks formed in the PME system help establish and maintain the "system of sponsorship," by which high-ranking officers influence the careers of promising young officers by requesting their assignment to their own staffs or recommending them for appropriate posts.[50] According to this theory, the contacts formed among these groups of peers and superiors eventually become the dominant influence on an officer's career.[51]

---

Gwendolyn Stevens, Fred Rosa, Jr., and Sheldon Gardner, "Military Academies as Instruments of Value Change." In fact, the evidence shows that this is largely a myth. Most three- and four-star officers in the U.S. military are not "ring knockers" but instead rise through the ranks from ROTC. This point was suggested to me by Paul Godwin.

[46]Peck, "Assessing the Career Mobility," p. 219.

[47]Janowitz, *The Professional Soldier*, p. 127.

[48]Segal, "Selective Promotion," p. 206. Since the PLA is moving in the opposite direction, from complete civilian penetration to minimal civilian penetration, perhaps the exclusiveness of academy attendance provides a necessary fire wall between the two groups.

[49]See David Moore and B. Thomas Trout, "Military Advancement: The Visibility Theory of Promotion," *American Political Science Review*, Vol. 72, 1978, pp. 452–468.

[50]Peck, "Assessing the Career Mobility," p. 219.

[51]Ibid., p. 220. For example, front-page scandals like the Tailhook incident resulted in few if any dismissals in the upper reaches of the U.S. Navy, as top brass circled the wagons around their peers.

Promotion based on personal ties, however, is nothing new in the Chinese army. Thus, the development of an advanced military education system in China presents a curious paradox. On the one hand, the expansion of the system is increasing the professionalism of the officer corps through the dissemination of standardized military knowledge and development of corporate unity. On the other hand, corporate unity might be undermining professionalism with the development of a new institutionalized source of personalistic networks. On balance, however, there is an important difference between these networks and past forms of factionalism, a difference that preserves the net benefit of the current phase of PLA professionalization. Specifically, the level of quality of the officers entering the system has fundamentally changed. Whereas previous generations of officers entered the system with distinctly nonprofessional entry standards (class background, ideological orthodoxy, etc.), the education- and testing-based entrance requirements for the current system ensure that the new officer corps meets a high minimum standard of competence. This meritocratic "floor" is a potential explanation for the relative professionalism of the U.S. (and future Chinese) officer corps in the face of a sometimes nonmeritocratic system of higher promotion.

In sum, the reform of the military education system and increasing pervasiveness of university education among the officer corps are net pluses for the professionalization of the PLA. While common PME attendance may create new sources of affiliations within the corps, these negative effects are offset by the higher standard of competence demanded by the system. Furthermore, changes in curricula demonstrate that the nature of military education in China has become more technical and apolitical in focus, with norms of professionalism being inculcated in its graduates.

## FUNCTIONAL SPECIALIZATION

Max Weber was among the first scholars to emphasize the importance of the "specialized division of labor" to bureaucracy. By extension, the modern army is a bureaucratic army, just as the modern nation-state is a bureaucratic state.[52] Perlmutter, Huntington, and other theorists of professionalism echo Weber, asserting that Western militaries have developed into complex, differentiated bureaucratic organizations, as dictated by the needs of modern warfare.[53] Modern warfare creates a particularly acute need for specialization and division of labor as the complexity and level of technological sophistication on the present-day battlefield can no longer be mastered by those Timothy Colton terms "universalists," officers characterized by their general knowledge of a number of subjects but no particular expertise.[54]

Historically, professional specialist officers skilled in "modern" technical arts replaced the part-time "amateur" officers of the 18th century aristocracy who relied on their intuitive "natural genius." This transition was first embodied in the Prussian General Staff of the late 19th century.[55] As a result of Prussian successes on the battlefield, the remaining nations of the world quickly followed suit. Even the leadership of the Soviet Red Army concluded by 1942 that egalitarian concerns in military hierarchy had to be sacrificed to specialized responsibility.[56] This type of functional specialization is indicative of professionalism because it buttresses the notions of both "expertise" and "corporateness." Functional specialization supports expertise in a different way than does education because it refers to specific technical competencies rather than basic educational knowledge. At the same time, it adds to overall military corporate-

---

[52]See Max Weber, *Economy and Society: An Outline of Interpretive Sociology*, pp. 114–130.

[53]See Amos Perlmutter, *The Military and Politics in Modern Times*, New Haven: Yale University Press, 1977, p. 24.

[54]See Timothy Colton, *Commissars, Commanders, and Civilian Authority: The Structure of Soviet Military Politics*, Cambridge, MA: Harvard University Press, 1979, p. 105.

[55]See Walter Goerlitz, *The History of the German General Staff: 1657–1945*, New York: Praeger Publishers, 1953.

[56]Jencks, *From Muskets to Missiles*, p. 18. Perlmutter and LeoGrande have persuasively argued that the Soviet Army, which was "converted by Lenin and Trotsky early in its evolution, followed the pattern of the classic European professional standing army." See Amos Perlmutter and William LeoGrande, "The Party in Uniform: Toward a Theory of Civil-Military Relations in Communist Systems," *American Political Science Review*, Vol. 76, December 1982, p. 785.

ness through the creation of subprofessions within the context of the brotherhood of military officers.

In the Chinese case, however, the historical dynamic posed ideology against professionalism. During the early Maoist era, the organizational ideal for cadres was to be both "red and expert" (*you hong you zhuan*).[57] In the 1950s, the supporters of Peng Dehuai tried to maintain the delicate equilibrium between the "one-man command" professionalism of the Soviet Red Army and the ideological focus of the Chinese military of the pre-Liberation period, though there is some debate as to whether Peng Dehuai favored professionalism or simply modernization.[58] After the purge of Peng in 1959 and Luo Ruiqing in 1965, the line between ideological (correct) and professional (incorrect) military cadres in the PLA was starkly drawn to forestall the development of what Mao derisively called the "purely military viewpoint," the belief that "military affairs and politics are opposed to each other."[59] As a result, the Chinese military in the mid 1960s, under the leadership of Lin Biao, became the power base used by Mao to launch the Great Proletarian Cultural Revolution against his perceived "revisionist" enemies in the CCP. The "little red book," which had originated in the military political system, was distributed nationwide, and the model soldier Lei Feng was offered for emulation by civilian and soldier alike. While the army's later suppression of the Cultural Revolution and the ignominious death of Lin Biao sullied the PLA's ideological status, it was still a largely ideological institution at the time of Mao's death in 1976. Indeed, more than half of its officer corps had been recruited and promoted

---

[57]Mao first referred to the concept of "red and expert" at the Third Plenum of the Eighth Party Central Committee in October 1957. See Mao Zedong, "Be Activists in Promoting the Revolution," *Selected Works of Mao Zedong*, Vol. 5, Beijing: Foreign Language Press, 1977, p. 489. This criterion was officially adopted by the CMC three years later: "To strengthen efforts to make our army more modern and revolutionary, including political and ideological work, we must first build a cadre contingent that is both red and expert." See Dong Lisheng, "The Cadre Management System of the Chinese People's Liberation Army (I)," p. 48.

[58]For an elaboration of this debate, see Zhu Fang, "Party-Army Relations in Maoist China, 1949–76," unpublished dissertation, Columbia University, 1994.

[59]Mao first referred to the "purely military viewpoint" at the Gutian Conference in 1929. See "On Correcting Mistaken Ideas in the Party," *Selected Works of Mao Zedong*, Vol. 1, pp. 105–108.

during those tumultuous years and were closely associated with the nonprofessional ethos of the late Maoist era.

In the post-Mao military, by contrast, an increasingly large percentage of the officer corps are functional specialists, reflecting the technocratic transformation reshaping the entire Chinese system.[60]  In this way, the PLA has begun to more closely resemble the post-purge Red Army of the Soviet Union, which actively sought to be a professional army rather than an "armed body for carrying out the political tasks of the revolution."[61]  Indeed, the Soviet officer corps was characterized by a "high degree of expertise, *specialized responsibility* and corporativeness" (emphasis added).[62]  While the military reformers in the PRC did not use their Russian counterparts as a model for emulation, the analogy is a useful one since the Soviet military more closely resembles the post-Mao PLA than Western militaries.

Despite this sweeping change, some still insist the post-Mao PLA has maintained and, in some cases, even strengthened the fusion between military and political roles in the officer corps.  In 1985, Yu Qiuli, who was then Director of the General Political Department, asserted:

> The reason we emphasize attaching importance to the political quality of the students [of the National Defense University] is, in accordance with the party's consistent policy, to uphold the unity of red and expert by developing in a comprehensive manner talented persons morally, intellectually, and physically.  We do not only stress politics, and not stress other things; *we also do not indulge in empty talk about politics separate from military affairs and specialized skills* [emphasis added].[63]

---

[60]Li and Bachman, "Localism, Elitism, and Immobilism," p. 89.

[61]Jencks, *From Muskets to Missiles*, p. 19.

[62]Ibid.

[63]Xinhua, 26 February 1986.

Indeed, one of the explicit purposes of founding the NDU was to train officers to be "all-around talents in military, political, and logistics work."[64] As a result, we should expect to see greater cross-fertilization in the career paths of the officer corps, especially between military and political postings.

However, the available data suggest the opposite conclusion. Table 2.3 shows the breakdown in career patterns and experience among the officer corps, which strongly suggests a trend towards functional specialization rather than cross-fertilization.

The table reveals that the number of officers engaged in military or political work *exclusively* (see "military only" and "political only") has risen, while the number of officers with cross-specializations (combinations like "military and political") has declined. Specifically, the percentage of PLA officers involved in purely military work rose from 33 percent in 1989 to 41 percent in 1994, while the distribution of officers engaged only in political work rose from 27 percent to 42

### Table 2.3

### Career Patterns and Experience

| Career Experience | 1989 | | 1994 | | Percent Change |
|---|---|---|---|---|---|
| | No. | Percent | No. | Percent | |
| Military only | 74 | 33 | 75 | 42 | +24 |
| Political only | 60 | 27 | 75 | 42 | +56 |
| Technical only[a] | 13 | 6 | 9 | 5 | −17 |
| Military and political | 46 | 21 | 13 | 7 | −62 |
| Military and technical | 15 | 7 | 3 | 2 | −71 |
| Political and technical | 9 | 4 | 3 | 2 | −50 |
| Military, political, and technical | 7 | 3 | 1 | 0 | −100 |
| Total | 224 | 100 | 179 | 100 | − |

SOURCE: Mulvenon PLA database.

[a]"Technical" refers to military personnel who are really only engineers or technicians in uniform, such as scientists under the Commission for Science, Technology, and Industry for National Defense (COSTIND). Nie Li is a good example of this type of officer.

---

[64]Luo Tongsong, Wang Jin, and Gai Yumin, "Leaders Attend Founding of Defense University," p. K6; and "Effectively Run the Defense University To Train Personnel Well-Versed in Advanced Modern Military Affairs," Xinhua Domestic Service, 18 December 1985, in FBIS, 19 December 1985, pp. K2–K5.

percent.[65]  Most striking is the drop in officers whose career paths
had involved both military and political work, which fell from 21 percent in 1989 to 8 percent in 1994.  These results contradict the pre-
vailing assumptions outlined above and suggest that the dominant
trend in military personnel patterns is division of labor, not politi-
cal/military fusion.

If we break these numbers down by age group, the transformation of
career patterns in the PLA is even more striking.  Tables 2.4a
and 2.4b show that younger officers (aged 45 to 57) in both the 1989
and 1994 pools are much more likely to be functional specialists in
military work than are older officers, although the gap between the
two age cohorts is narrowing in favor of functional specialization
across the entire officer corps.  Between 1989 and 1994, the percentage of officers aged 45 to 57 with purely military backgrounds held
steady around 50 percent.  In contrast, 21 percent of officers aged 64
to 75 had followed "military only" careers, a number that increased
to 36 percent in 1994.  The same holds true for midrange officers
aged 58 to 63, whose percentage of "military only" careers increased
52 percent from 29 percent in 1989 to 44 percent in 1994.  Thus, while
a greater percentage of younger officers still have a purely military
background, the number of midrange and older officers devoted
solely to military work has also increased significantly since 1989,
due to the aging of the more specialized 1989 cohort.

The same trend can be seen in exclusively political careers, but the
age distribution is the exact opposite.  Younger officers are still less
likely to pursue exclusively political careers than are their older
counterparts, despite the fact that the overall number of officers
whose careers involve only political work has risen across the board
between 1989 and 1994.  In 1989, 34 percent of officers aged 64 to 75
had followed exclusively political careers, as compared with 22 percent of officers aged 45 to 57.  In 1994, the difference between the two
age cohorts narrowed at the same time that the overall level of spe-

---

[65]While the percentage of officers engaged in technical work alone fell by 17 percent,
this decline was primarily a function of the small number of officers in that category.

Table 2.4a

Career Patterns and Experience in the Officer Corps, by Age Group

| | 1989 | | | | 1994 | | | |
|---|---|---|---|---|---|---|---|---|
| Career Path | 45–57 No. (%) | 58–63 No. (%) | 64–75 No. (%) | All No. (%) | 45–57 No. (%) | 58–63 No. (%) | 64–75 No. (%) | All No. (%) |
| Military only | 30 (51) | 33 (29) | 11 (21) | 74 (33) | 16 (50) | 36 (44) | 23 (36) | 75 (42) |
| Political only | 13 (22) | 29 (26) | 18 (34) | 60 (27) | 12 (38) | 34 (41) | 29 (45) | 75 (42) |
| Technical only | 4 (7) | 7 (6) | 2 (4) | 13 (6) | 1 (3) | 5 (6) | 3 (5) | 9 (5) |
| Military and political | 6 (10) | 29 (26) | 11 (21) | 46 (21) | 0 (0) | 6 (7) | 7 (11) | 13 (7) |
| Military and technical | 2 (3) | 8 (7) | 5 (9) | 15 (7) | 2 (6) | 1 (1) | 0 (0) | 3 (2) |
| Political and technical | 2 (3) | 5 (4) | 2 (4) | 9 (4) | 1 (3) | 1 (1) | 1 (1.5) | 3 (2) |
| Military, political, and technical | 2 (3) | 1 (1) | 4 (8) | 7 (3) | 0 (0) | 0 (0) | 1 (1.5) | 1 (.5) |
| Total | 59 (100) | 112 (100) | 53 (100) | 224 (100) | 32 (100) | 83 (100) | 64 (100) | 179 (100) |

SOURCE: Mulvenon PLA database.

Table 2.4b

Percentage Change from 1989 to 1994

| Career Path | 45–57 | 58–63 | 64–75 | All |
|---|---|---|---|---|
| Military only | –2 | +52 | +71 | +27 |
| Political only | +73 | +58 | +32 | +67 |
| Technical only | –57 | 0 | +25 | –17 |
| Military and Political | –100 | –73 | –48 | –67 |
| Military and technical | +100 | –86 | –100 | –71 |
| Political and technical | 0 | –75 | –63 | –50 |
| Military, political, and technical | –100 | –100 | –81 | –83 |

SOURCE: Mulvenon PLA database.

cialization grew more than 67 percent. Specifically, older political officers outnumbered their younger counterparts in 1994 at a rate of 45 percent to 38 percent, leaving only a 7 percent gap between the two age groups. Thus, we see the same "leveling" effect in political work that was evident in military work, suggesting that the percentage of officers pursuing functionally specialized careers is increasing steadily throughout the officer corps.

Cross-fertilization among military, political, or technical positions, on the other hand, seems an increasingly rare career path in the PLA officer corps. As seen in Tables 2.4a and 2.4b, the percentage of officers whose careers involved multiple specializations dropped precipitously between 1989 and 1994, in some cases disappearing altogether. Military-political careers dropped 67 percent, military-technical 71 percent, political-technical 50 percent, and military-political-technical careers fell over 83 percent. Looking at officers with military-political backgrounds, the importance of the age variable is especially apparent. In 1989, older officers whose careers had involved both military and political work outnumbered younger officers with similar backgrounds by a margin of 21 percent to 10 percent. This distribution reached an extreme point in 1994, when 11 percent of officers aged 64 to 75 had military-political experience and not a single younger officer had such a background. These results strongly suggest that the PLA as a military organization is abandoning generalism and cross-fertilization in favor of functional specialization, as dictated by the demands of modern technological warfare.

To understand the fundamental nature of this shift, it is useful to examine the career histories of two members of the senior officers corps. Liu Huaqing (aged 78 in 1994), vice-chairman of the Central Military Commission, is the lone surviving senior PLA leader with a combined military, political, and technical background. He has served as political commissar of a military area command, commander of a service branch, and director of the PLA's research and development organizations, thus covering all three major specializations in the Chinese military. Although he is by no means a "generalist" opponent of functional specialization and professionalization (in fact, he is at the forefront of the PLA's modernization program), his career path is a quickly vanishing phenomenon.[66]  In contrast, Fu Bingyue (aged 54 in 1994) represents the new breed of functional specialist in the officer corps.  He has successively served as platoon leader, staff officer, company commander, deputy battalion commander, regiment commander, deputy division commander, division commander, corps commander, deputy commander of a military region, and finally commander of a military region.  This path is very common among his peers, only 12 percent of whom have deviated from specialized careers.

In sum, the data strongly suggest that the PLA officer corps is becoming increasingly specialized among military, political, and technical careers.  On average, the percentage of officers engaged in specialized careers rose 26 percent while those with cross-fertilized careers fell more than 68 percent.  Younger officers are only more specialized than older officers in the case of purely military careers, which is appropriate considering that military officers, by definition, should be more "professional" than their political counterparts.  Overall, however, the rate of functional specialization is increasing regardless of age group and suggests that the PLA is continuing to professionalize itself to meet the demands of modern high-tech warfare.

## RETIREMENT NORMS

As Melanie Manion argues, "personnel changes in the party and government of communist systems are typically politicized and personalistic," occurring as a result of natural death, political error, or "consolidation of personal power from the top."[67]  Cadres who are able to deftly maneuver through this type of system often enjoy "*de facto* lifelong tenure."  In the Chinese political arena, adherence to these rules of the game resulted, by the late 1970s, in a large, calcified stratum of veteran cadres monopolizing leadership at all levels of the government and party.  Many of these veterans traced their revolutionary lineage to the earliest days of the party but were increasingly

---

[66]It could be argued that Liu Huaqing is a special case because of his genuine status as a member of the revolutionary generation.  For additional examples of military-political officers from the 1989 pool, examine the biographies of Li Yuan, Jing Demin, Liu Cunxin, and Zhang Zhenxian.

[67]Melanie Manion, "Politics and Policy in Post-Mao Cadre Retirement," p. 1.

impeded by their "generally low levels of education, expertise, and sheer physical and mental vigor."[68]  In 1978, the government introduced new retirement regulations to break up this bureaucratic logjam, paving the way for better-qualified, younger cadres to manage the drive for economic growth and modernization.  As Manion points out, however, this policy was not simply an ad hoc measure to deal with a temporary crisis.  Instead, the long-term goal was the "regularization or systemization of retirement," which would provide an enforceable norm for future age-based exits from the political scene.[69]

Similar serious problems of leadership mobility have existed in the PLA since 1949.  For many years after the founding of the People's Republic, the revolutionary generation of military/political leaders stubbornly held onto their institutional power, preventing their "formal" successors from exercising legitimate authority.  At the time of Mao's death in 1976, a significant number of these founding military leaders were still alive and taking part in military work.  The result was a calcification of the military leadership at every level of the system, creating what Harlan Jencks, writing in 1984, called a "generation gap."[70]  Although a senior military leader himself, Deng Xiaoping was especially cognizant of the need to correct this situation, as seen in his March 1980 comments to the CMC:

> We must particularly take note of the ages of the cadres at the military region, army and division levels.  All of them are pretty much the same age—rather old.  In a few years, they will all be elderly. Not only will they be unable to work at the army or divisional level, but they'll find it difficult to work in the military region commands or the general headquarters.  This is a matter of a law of nature. How old will you comrades here be in five years?  I'm afraid most of you, though not all, will find it hard to keep on working.  Seven or eight years from now, you'll be past 70.  How could you see things on the battlefield? If war should really break out, could you fight for three days and nights without sleep? The current move to reduce bloatedness will also help to renew the ranks of our cadres.  The reason the lower-level cadres could not be promoted is that older

---

[68]Ibid., p. 2.

[69]Ibid., pp. 2–3.

[70]Jencks, *From Muskets to Missiles,* p. 1.

ones have stood in their way. . . . Today key posts can only be held
by veteran cadres—in a few years we'll really be in a fix![71]

Despite these early efforts, however, Deng encountered great
difficulty uprooting his elderly comrades. This can be explained by
two factors. First, many of the original generation of officers
continued to maintain powerful patronage networks throughout the
army that ensured their influence and power in the system. This
situation was legitimized by a Chinese political culture that generally
favors "moralized and personalistic authority relations" over
institutions and legal rationality, and a Leninist structure that
provided no regularized avenues of elite-level leadership succession
except through purge.[72] Second, the political and social chaos of the
Cultural Revolution had permitted certain military leaders to build
strong power bases in China's regional military units. A few of these
leaders, such as Wei Guoqing and Xu Shiyou, had been instrumental
in engineering Deng's political comeback in the late 1970s and
strongly resisted surrendering their position through retirement.[73]

In the hopes of finally eliminating these pockets of resistance, the
CMC in the mid-1980s began a series of institutional reforms, including forced retirements of veteran cadres and organizational consolidations at all levels of the armed forces.[74] Deng Xiaoping argued
persuasively for the policy:

> One of the most important questions relating to structure and sys-
> tems is the establishment of a system of military service and of re-
> tirement for officers. In the fifties, regulations were drawn up to
> govern the military service of officers, but they didn't work and were

---

[71]Deng Xiaoping, "Streamline the Army and Raise Its Combat Effectiveness," 12 March
1980, in *Selected Works of Deng Xiaoping,* Beijing: Foreign Language Press, 1982, p.
270–271.

[72]See Lowell Dittmer, "Chinese Informal Politics," *The China Journal,* No. 34, July
1995, p. 1; Ian Wilson and You Ji, "Leadership by 'Lines': China's Unresolved
Succession"; and Eberhard Sandschneider, "Political Succession in the People's
Republic of China: Rule by Purge," *Asian Survey,* Vol. 25, No. 6, June 1985, p. 638.

[73]An excellent analysis of the PLA's resistance to Deng's reform program and military
personnel changes can be found in Richard Baum, *Burying Mao: Chinese Politics in the
Age of Deng Xiaoping,* Princeton: Princeton University Press, 1994, pp. 121–124.

[74]The most publicly visible of these organizational consolidations was the reduction
in the number of military regions from 11 to 7. See Ibid., pp. 183–184.

later dropped. Actually, they were good on the whole and reflected the correct approach. . . . We must have a retirement system. . . . Since the army has to fight, the retirement age for military cadres should be lower than that for civilians. Of course, the regulations must be practicable. The vitality of our whole state will be affected if it fails to establish a retirement system. . . . With such a system, everyone will know when he is to retire, and the necessary arrangements can be made more easily. Otherwise, the problems have to be handled case by case and everything becomes difficult.[75]

However, the implementation of this new retirement policy was difficult and required careful and sensitive handling. First, veterans were understandably "reluctant to part with the benefits of high office which they had enjoyed for years," as well as the concomitant prestige.[76] Concerns over housing and benefits packages had to be adequately addressed as well as the question of younger candidates for promotion.[77] Secondly, the reformers had to be careful not to antagonize elder commanders who still presided over vast patronage networks vital to the legitimacy of the Dengist regime.

While the initial resistance from veteran officers was expectedly strong, by 1986 moderate success had been achieved in uprooting recalcitrant officers from their positions, primarily because the military leadership had taken great pains to soften the blow with generous pensions, retention of perquisites, and other "golden parachutes." To serve as an example to their subordinates, 40 senior officers at or above the corps level retired in late December 1984 in order to "give way to younger and more competent leaders."[78] In March 1985, it was announced that 47,000 officers would be retired by the end of 1986,[79] and an additional 70,000 to 80,000 officers

---

[75]Deng Xiaoping, "Streamlining the Army and Raise Its Combat Effectiveness," 12 March 1980, in *Selected Works*, p. 273.

[76]Joffe, *The Chinese Army After Mao*, p. 131.

[77]"Elderly Veteran Officers Retire From PLA," Xinhua, 5 March 1985, in FBIS, 6 March 1985, p. Kl.

[78]"Senior PLA Officers Retire From Leading Posts," Xinhua, 29 December 1984, p. K2. Note: The corps level was replaced by the "group army" during the 1980s.

[79]Ibid., p. Kl.

would be demobilized by 1990.[80] These large groups of officers had joined the PLA before 1949 and were of relatively junior rank in lower-level units. Dispelling the question of double standards or "stratification" (to use Manion's term for the policy of leadership exemption), the 6 highest-ranking officers in the army also retired en bloc from the Politburo in September 1985, providing a leading example to their veteran comrades in the middle and upper levels of the military leadership.[81]   Table 2.5 shows the overall impact of this first series of retirements, as the average age of military region, group-army, division- and regiment-level officers declined significantly from 1982 to 1986, dropping an average of 10 percent.[82]

Table 2.5

Average Ages of Military Elites, by Level

|  | 1982 | 1986 | Change |
|---|---|---|---|
| Military Region and above | 65.3 | 57.1 | −8.2 yrs |
| Army | 56.8 | 49.6 | −7.2 yrs |
| Division | 48.3 | 43.5 | −4.8 yrs |
| Regiment | 39.1 | 37.2 | −1.9 yrs |

SOURCE: BG Michael Byrnes, "The Death of a People's Army," p. 134. Used by permission.

---

[80]Baum, *Burying Mao*, p. 184. At a February 1985 meeting of retiring Beijing Military Region cadres, General Qin Jiwei thanked them for their sacrifice with a verse from an ancient poem: "Knowing that the sun would set soon, the old ox ran without being whipped." See Zhao Su, "Qin Jiwei Praises, Encourages, Retired PLA Cadres," Xinhua Domestic Service, 12 February 1985, in JPRS-CPS-85-020, 4 March 1985. See also Yitzhak Shichor, "Demobilization: The Dialectics of PLA Troop Reduction," *China Quarterly*, No. 146, June 1996, pp. 336–359.

[81]The officers who retired were Ye Jianying, Nie Rongzhen, Xu Xiangqian, Wang Zhen, Song Renqiong, and Li Desheng. Of course, the political impact of these retirements was mitigated by the fact that all six officers were appointed to the Central Advisory Commission, where they continued to exercise high-level influence. Li Desheng also remained very influential in his new posting as political commissar of the NDU. See Baum, *Burying Mao*, pp. 184–185.

[82]Additionally, Michael Byrnes states that by the mid-1980s, the average age of brigade, division, and GA commanders (38, 45, 51) was equal to or less than comparable numbers for Western forces. For both tables, see Michael Byrnes, "The Death of a People's Army," in George Hicks, ed., *The Broken Mirror: China After Tiananmen*, Chicago: St. James Press, 1990, p. 134. This change was also verified in an official announcement in 1985. See Xinhua, 5 July 1985.

While some of these reductions can be attributed to the civilian redesignation of some uniformed positions, such as officers in health care, education, and research, these numbers strongly suggest that the retirement program was not only successful but pervasive.[83]

The relatively long period of time it took to retire these veteran officers was largely a function of the fact that in the late 1970s and early 1980s there were no formal regulations on the books for age-based retirement from the military. As Ellis Joffe argues, the military leadership could only "coax rather than coerce aging officers to leave the armed forces."[84] Once this first group was demobilized, however, the CMC began to pursue the longer-term goal: implementation of an enforceable, age-based retirement norm. Indeed, there was a serious concern that the entire situation would simply repeat itself if the CMC did not take swift and resolute action. To this end, new regulations on military cadre retirement were promulgated in the 1984 General Political Department circular, "Seven-Year Plan on the Building of Leadership Groups and the Four Transformations of the Cadre Contingent."[85] This general set of principles was given more specificity in 1988, when the National People's Congress passed a bill entitled "Regulations for Military Service of Active Duty Officers of the Chinese People's Liberation Army."[86] The 1988 regulations codified age limits for various positions in the military hierarchy and dictated the retirement of officers who had served more than 30 years or had reached the age of 50 without attaining a certain rank. They provided extensions in exceptional cases, but these were not to exceed 5 years. At the same time, these reforms had limits: CMC members and principal leaders of the three General Departments were not subject to the rules.

In 1994, the National People's Congress promulgated the "Decision on Revising the Service Regulations Governing Active Duty Officers in the PLA," which slightly lowered the previous age limits and im-

---

[83]Dreyer, "The New Officer Corps," p. 316.

[84]Joffe, *The Chinese Army After Mao*, p. 131.

[85]Dong Lisheng, "The Cadre Management System of the Chinese People's Liberation Army (I)," pp. 59–60.

[86]"Regulations for Military Service of Active Duty Officers of the Chinese People's Liberation Army," Xinhua, in FBIS, 8 September 1988, pp. 37–41.

posed new limits on the ages of CMC members and heads of the three General Departments. Henceforth, directors of general departments, commanders of military regions and the Beijing Garrison, and commandants of military schools had to retire at age 65 while their deputies must leave their post by age 63.[87] If required by work, and with the recommendation of the CMC and the approval of the CCP Central Committee, the retirement limit for general department personnel could be extended to 68 and commanders to 66. Commanders of provincial military districts, group armies, other garrisons, and regional military schools must retire at 60, and their deputies must retire at 58 years of age. The retirement limit for these principal officers can be extended to 62. This pattern of age-based retirement continues down to the platoon level, where personnel can be demobilized at age 30 (see Table 2.6).[88]

However, the mere existence of regulations does not guarantee their enforcement, especially in the case of high-level personnel changes. Using U.S. government sources, it is possible to follow the career paths of the officers present in the 1989 pool who are missing in the 1994 pool and determine the age and position at which they were retired.[89] Careful examination of this group suggests the PLA is

---

[87]"Active Service Regulations Governing Active Duty Officers of the People's Liberation Army," pp. 37–38. See also Chang Hong, "NPC Plans to Reshuffle Top Military Rankings," *China Daily*, 6 May 1994, p. 1, in FBIS, 6 May 1994, pp. 29–30; Willy Wo-Lap Lam, "Reasons for Military Reshuffling Examined," *South China Morning Post*, 5 September 1995, p. 1, in FBIS, 6 September 1995, p. 33; Willy Wo-Lap Lam, "Chengdu, Jinan Military Region Leaders Retire," *South China Morning Post*, 17 December 1994, p. 10, in FBIS, 19 December 1994, pp. 31–32; Yoshiaki Hara, "PLA Reshuffles Leaders Under Retirement Age System," *Yomiuri Shimbun*, 27 November 1994, p. 5, in FBIS, 28 November 1994, p. 45; and Bruce Gilley, "Air Force Chief's Retirement Goes Unreported," *Eastern Express*, 29 November 1994, p. 8, in FBIS 29 November 1994, pp. 33–34.

[88]The primary source for this table is "Active Service Regulations Governing Active Duty Officers of the People's Liberation Army," pp. 37–38. For details on the CMC members, see Yueh Shan, "Fifty-Eight Generals to Retire," *Cheng Ming*, No. 207, 1 January 1995, pp. 18–19, in FBIS, 16 February 1995, pp. 42–43.

[89]This unclassified source is the annual *Directory of P.R.C. Military Personalities*, Hong Kong: Defense Liaison Office, U.S. Consulate General, and includes every issue from 1989 to 1994. This source delineates the Chinese military hierarchy down to the regimental level, although age data are only available from the *Who's Who* editions.

rapidly approaching an operational retirement norm.[90] First, the raw data reveal that the membership consistency from the 1989 pool to the 1994 pool was 53 percent; that is, 53 percent of the military leadership in 1994 was also present in the 1989 pool. On a macrolevel, the average age of officers in both pools was 57.27 in 1989, while the average age of those officers absent in the 1994 pool (presumably from retirement) was 62.75 in 1989 (see Table 2.7).[91]

### Table 2.6

### Retirement Ages and Extensions for Active Duty Officers

| Officer Billet | Max. Exit Age | Max. Extension |
|---|---|---|
| Platoon Commander/Commissar | 30 | – |
| Company Commander/Commissar | 35 | – |
| Battalion Commander/Commissar | 40 | – |
| Regiment Commander/Commissar | 45 | – |
| Division Commander/Commissar | 50 | 5 years |
| Army Commander/Commissar | 55 | 5 years |
| MD, Group Army, and Garrison Deputy | 58 | NA[a] |
| MD, Group Army, and Garrison Principal | 60 | 62 years old |
| Military Region Deputy Commander/Commissar | 63 | NA |
| Military Region Commander/Commissar | 65 | 66 years old |
| General Department Dep. Director/Commissar | 63 | NA |
| General Department Director/Commissar | 65 | 68 years old |
| Military School Commandant | 65 | NA |
| CMC Member | 70 | 72 years old |
| CMC Chairman/Vice-Chairs | No limit | No limit |

SOURCE: "Active Service Regulations Governing Active Duty Officers of the People's Liberation Army," pp. 37–38.

[a]NA = Not applicable.

---

[90]On an ironic note, the entire personnel reform is being presided over by a septuagenarian (Liu Huaqing) and octogenarian (Zhang Zhen) brought out of retirement to retire younger officers.

[91]It must be noted, however, that the percentage of PLA officers in the 65–69 age group *rose* from 14.3 percent to 21 percent from 1989 to 1994. Excepting the small number of officers in the CMC and the General Departments, this somewhat contradictory result suggests that "informal politics" continue to play a small but significant part in military advancement or that many officers chosen for high position are retained for a few years beyond the mandatory retirement age in order to solidify factional balances within the PLA or because of their outstanding qualifications.

Table 2.7

Comparison of Average Ages in 1989 Pool

| Officer Pools | Average Age in 1989 |
|---|---|
| Officers retired after 1989 | 62.75 |
| Officers in both 1989 and 1994 pools | 57.27 |

SOURCE: Mulvenon PLA database.

This strongly suggests adherence to a retirement norm, since nearly all of the latter group of officers, on average, would have reached the age of 65 over the 6-year period between 1989 and 1994, while the former group of officers, on average, would still be eligible for high-level postings.

If we analyze the data on the 84 retired officers by the position they held when they were removed from service (see Table 2.8), we see even stronger evidence for the opening phase of a functioning retirement norm.

While the average age of retirement for each position is still higher than the prescribed maximum (due primarily to final weeding of older officers from the ranks), the average age of the subsequent re-

Table 2.8

Average Age of Officers Retired from the 1989 Pool over the Period
1989–1994 and Their Replacements, by Position

| Position | No. | Age Retired | Max. Extension | Replacement | Difference |
|---|---|---|---|---|---|
| CMC Vice-Chair | 1 | 85 | No Limit | 76 | −9 yrs |
| CMC Member | 3 | 75.3 | 70/72 | 62.3 | −13 yrs |
| Military School Principal | 8 | 67.4 | 65 | 64 | −3.4 yrs |
| General Dept. Principal | 1 | 72 | 65/68 | 61 | −9 yrs |
| MR Principal | 8 | 68.75 | 65/66 | 59 | −9.75 yrs |
| General Dept. Deputy | 3 | 72.3 | 63 | 54 | −18.3 yrs |
| MR Deputy | 10 | 65.6 | 63 | 59.1 | −6.5 yrs |
| MD Principal | 1 | 68 | 60/62 | NA[a] | NA |
| MD Deputy | 4 | 65 | 58 | 50 | −15 yrs |

SOURCE: Mulvenon PLA database.

[a]NA = Not applicable.

placement shows a striking decrease. For example, the average age of a member of the CMC, China's highest military decisionmaking body, dropped from 75.3 to 62.3 years. Equally striking, the average age of the important military region principals (commander and political commissar) dropped from 68.75 to 59 years. The average ages of these new officers are much more in line with the CMC's vision of age-based retirement since their average ages are slightly less than 6 years *below* the prescribed maximum for that position. Their predecessors, by contrast, were on average slightly less than 6 years *above* the prescribed maximum. This means that the replacement officers can serve out the minimum term for a military region principal (3 to 5 years according to the "1994 Regulations for Active Duty Officers") without violating the maximum retirement age.[92]

Given these findings, can we conclude that there is a functioning retirement norm in the PLA? As Manion points out, the existence of a norm is not an all-or-nothing matter but a matter of degree.[93] Yet if we begin with the assumption of "de facto lifelong tenure" as the status quo ante, then the reforms of the 1980s and 1990s have indeed transformed the nature of the PLA personnel system. Except for the very highest reaches of the leadership, such as vice-chairmen of the CMC, it appears that military officers no longer hold their postings indefinitely and have come to expect regular enforcement of exit from the armed forces based upon their age and position. Additionally, their replacements are far younger in age, significantly lowering the overall age of the military leadership and setting in place the foundation for a permanent rotation. Such orderly turnover, which was unthinkable during the Mao and early Deng eras, now appears to be the norm in the PLA and marks a major step forward in its professionalization.[94]

---

[92]See "Active Service Regulations Governing Active Duty Officers of the People's Liberation Army," p. 37.

[93]Manion, *Retirement of Revolutionaries in China*, p. 153.

[94]These changes in the retirement patterns within the officer corps are good indicators of the level of overall professionalization since they generally reveal the level of bureaucratic "rationality" within the system (i.e., the relation between formal position and informal power) and, more importantly, the extent to which the officer corps is "obedient" to the state, which is a key dimension of Huntington's vision of pure professionalism. In his ideal-type "professional" military, promotions and

As a postscript, the now annual personnel changes in 1995 and 1996 strongly confirm this trend toward age-based retirement. In December 1995, deputy chief of the General Staff Li Jing (65), political commissar of COSTIND Dai Xuejiang (65), political commissar of the General Logistics Department Zhou Keyu (66), as well as military region commanders Li Xilin (65), Gu Hui (66), and Cao Pengsheng (66) were all retired at the prescribed age for their position.[95] Likewise, December 1996 witnessed the retirements of People's Liberation Army Navy (PLAN) Commander Zhang Lianzhong (65), PLAAF Commander Yu Zhenwu (65), COSTIND Director Ding Henggao (65), Jinan MR Commander Zhang Taiheng (65), and Beijing MR Political Commissar Gu Shanqing (65).[96] Such regularized turnover would have been unthinkable in the prereform PLA.

---

dismissals are based on routinized meritocratic principles rather than personalistic affiliation, and retirement is based upon universally applicable criteria.

[95]For further information on recent reshuffles, see Chang Hsiu-fen, "Major Reshuffle of China's Military Hierarchy—Fourth Generation of Military Officers Take Over Important Posts," *Kuang Chiao Ching,* 16 September 1995, No. 276, pp. 16–18, in FBIS, 29 September 1995, pp. 31–32; and Willy Wo-Lap Lam, "Reasons for Military Reshuffling Examined," *South China Morning Post,* 5 September 1995, p. 1, in FBIS, 6 September 1995, pp. 32–33. The Chinese leadership has reshuffled top military positions in November–December in each of the last three years. Official reports indicate that 1997 will continue the pattern. See Willy Wo-Lap Lam, "Further Changes Planned," *South China Morning Post,* 7 October 1995, p. 8, in FBIS, 10 October 1995, pp. 46–47.

[96]See Willy Wo-Lap Lam, "Top PLA Posts Go To High-Tech Experts," *South China Morning Post,* 4 December 1996, p. 1; and Liu Ping-chun, "Jiang Zemin Orders Retirement of Five Navy, Air Force Generals," *Ming Pao,* 3 December 1996, p. A12.

# Chapter Three

# COHORT ANALYSIS

Cohort analysis identifies groups within the officer corps that, by virtue of common age or experience, share certain values or beliefs that can be differentiated. This can be contrasted with factional analysis, which is concerned with identifying the distribution of political power within the military leadership, particularly with reference to the dynamics of civil-military relations.[1] Most of the literature on PLA professionalism discusses factions, not cohorts, because the former were the dominant feature of civil-military relations in the Maoist and early Dengist period. With the death of the military elders and the rise of more professional warfighters, however, the factional dynamic within the officer corps is changing. The current senior officer corps operates in a much more complex bureaucratic environment, without referents of factional behavior such as a field-army system to guide its intramilitary strategic behavior. As a result, cleavages within the senior officer corps are increasingly policy- or interest-driven. This makes cohort analysis, with its emphasis on the diversity of values and beliefs, the more appropriate methodology for analyzing the two data pools.

For the purposes of this report, our definition of a cohort centers on the general concept of "group affiliation," which operates on a "we-they" distinction. Group affiliations within the officer corps include

---

[1]For the classic discussion of factions in Chinese politics, see Lucian Pye, *The Dynamics of Chinese Politics*, Cambridge: Oelgeschlager, Gunn, and Hain, 1981. An excellent discussion of factions in a Chinese military context circa 1992 can be found in Swaine, *The Military and Political Succession*, pp. 16–26.

affiliations with military and nonmilitary groups.[2]  These associations contribute to the development of corporatism because they strengthen the "organic unity" and self-identification of the officer corps.[3] Intramilitary affiliations can be both formal and informal. Formal in-service affiliations include relationships derived from the course of official duties, most importantly shared combat or staff experience. These types of ties are examined in a later part of the chapter, which compares the relative war experiences and the corps affiliations of the two pools of officers. Informal affiliations, on the other hand, include ties forged in quasimilitary functional associations and those developed during military education.  In order to evaluate these ties, we compare the shared educational attendance experiences of the two pools of officers. The second major category of affiliations is nonmilitary groups, which include preservice affiliations with a particular class or geographic area.[4] To this end, one part of this chapter is devoted to the examination of the birth origins of the senior officer corps.

This chapter is divided into two basic sections.  The first summarizes and evaluates the previous frameworks of factional and cohort cleavages within the PLA, in particular the "field-army thesis." The second introduces a potential alternative framework, the "generational thesis," and evaluates each of its key cohort variables in light of the two data pools.

## FIELD ARMIES

William Whitson and Huang Chen-hsia's *The Chinese High Command* is the strongest English-language presentation of the field-army thesis.  The five field armies were given formal designation in

---

[2]This discussion is drawn largely from Jencks, *From Muskets to Missiles*, p. 9.

[3]Ties between high-ranking military and party officials receive most of the attention in the literature, but ties between lower-level military and party officials are also extremely important, especially given recent regulations requiring localities to pay an increasing part of the burden for military units in their area.

[4]For example, most of the original military and party revolutionary generation originated in the central and upper Yangzi River valley.

early 1949 by the CCP.[5]  The field armies themselves were the final configuration of Communist military forces on the eve of Liberation and arguably provided the structure of regional military influence for decades to come.  Historically, the 1st FA controlled northwest China, the 2nd FA controlled the central and southwest, the 3rd FA controlled the east, the "southern" 4th FA controlled the southeast, the 4th FA controlled the northeast, and the 5th FA (or North China Field Army) controlled the north.  The theory's proponents believed that there was a high correlation between field army affiliation and career promotion patterns in the post-Liberation PLA.

For years, the field-army thesis has been the focus of intense debate within the field of China studies.  Perhaps the strongest criticism of the Whitson thesis can be found in William Parrish's statistical critique, which argues that bureaucratic rather than factional models are more appropriate for the subject.  Other less penetrating criticisms have questioned specific factional memberships or interactions.  While the advantages and disadvantages of this paradigm have been extensively discussed in other fora, it is clear that the dying off or retirement of the original revolutionary generation has weakened the potential explanatory power of this model.[6]  Nonetheless, it still receives a significant amount of attention, and Chinese interlocutors continue to stress its importance to foreign scholars.[7]

For comparison purposes, therefore, the field-army distribution of the 1989 and 1994 officer corps is displayed in Table 3.1.

The data in the table suggest that the 4th FA continues to dominate, although the strength of this group is diluted between the 4th FA and what is known as the southern 4th FA.[8]  If the 4th FA is divided along these lines, then the strongest single grouping is the 3rd FA, which makes up 28 percent of the total.  The drop of the 3rd FA from 38 percent in 1989 to 28 percent in 1994 is surprising given the promotion

---

[5]For the definitive history of China's field armies, see William Whitson and Huang Chen-hsia, *The Chinese High Command.*

[6]William Parrish, Jr., "Factions in Chinese Military Politics," *China Quarterly,* Vol. 56, October-December 1973, pp. 667–699.

[7]James Mulvenon, Personal interviews in Beijing, March 1997.

[8]During the Chinese Civil War period, the 4th FA split into two parts.  The 4th FA conquered northeast China, while the southern 4th FA fought in southeast China.

Table 3.1

Distribution by Field Army Origin

| FA | 1989 | | 1994 | |
| --- | --- | --- | --- | --- |
| | No. | Percent | No. | Percent |
| 1st | 3 | 3 | 3 | 3 |
| 2nd | 14 | 13 | 8 | 9 |
| 3rd | 39 | 38 | 24 | 28 |
| 4th | 43 | 41 | 39 | 45 |
| 5th | 5 | 5 | 9 | 10 |
| Unknown | — | — | 3 | 3 |
| Total | 104 | 100 | 86 | 100 |

SOURCE: Mulvenon PLA database.

in 1992 of one of its most important veterans, Zhang Zhen, to the vice-chairmanship of the CMC with the portfolio for personnel decisions.[9]  On the other hand, it could be hypothesized that the new core of Zhang's clique is officers who graduated from the NDU during his tenure as president and not necessarily officers tied to the 3rd FA.  This hypothesis will be evaluated later in the chapter.  Most striking, however, is the continued poor showing of Deng Xiaoping's 2nd FA, which dropped from a relatively low 13 percent in 1989 to 9 percent in 1994.  While many have argued that Deng maintained military control through a 2nd FA/3rd FA alliance, it is clear that the 2nd FA makes up an increasingly smaller component of that group.

## A NEW PARADIGM?

In response to the criticisms of the field-army thesis and the rapid disappearance of actual participants in the original factions, a variety of alternative paradigms have been put forward to strengthen its explanatory power.  This concept, which centers on the identification of age-based "military generations," has been recently elaborated by Michael Swaine.[10]  Swaine hypothesizes that China's military leadership can be divided into three basic generations: the geron-

---

[9]Some have asserted that Zhang Zhen is not identified within the PLA by his FA affiliation but by his presence on the Long March. This would mean that the faction around him would be based more on his revolutionary prestige than on his 3rd FA service.

[10]Swaine, "Generations in the PLA," unpublished paper.

tocratic elite, a middle generation of officers between the ages of 50 and 70, and a junior generation of officers between the ages of 35 and 50. These three generations are distinguished by each group's weltanshauung, which has been shaped by historical experiences unique to each group. In Swaine's opinion, the differences between these three generations "are probably far more important to political order in China today than more 'traditional' distinctions stemming, for example, from the contrasting functional duties of commanders and commissars."[11]

One fruitful method of quantifying the Whitson/Swaine notion of generations is to examine the dates of officer entry into the CCP and the PLA. Tables 3.2 and 3.3 separate the dates of entry along the faultlines of major periods in Party history.[12]

The data reveal that the bulk of the active military leadership (88 percent) falls into Swaine's second generation, which the author believes should be viewed as "primarily military figures" in contrast

Table 3.2

Date of Joining the Chinese Communist Party

| Years | 1989 | | 1994 | |
|---|---|---|---|---|
| | No. | Percent | No. | Percent |
| 1929–1937 | 19 | 9 | 2 | 1 |
| 1938–1944 | 76 | 34 | 14 | 8 |
| 1945–1949 | 86 | 38 | 86 | 48 |
| 1950–1954 | 22 | 10 | 40 | 22 |
| 1955–1965 | 6 | 3 | 30 | 17 |
| 1966–1980 | 0 | 0 | 4 | 2 |
| Unknown | 15 | 7 | 3 | 2 |
| Total | 224 | 100 | 179 | 100 |

SOURCE: Mulvenon PLA database.

---

[11]Whitson also suggested the importance of generations in the PLA, although it received much less attention than the pure field-army thesis. (I am indebted to Harlan Jencks for this reference.)

[12]The dates correspond to the following historical periods: 1929–1937 (pre-Anti-Japanese War), 1938–1944 (Anti-Japanese War), 1945–1949 (War of Liberation against the Kuomintang), 1950–1954 (Korean War), 1955–1965 (pre–Cultural Revolution), 1966–1994 (Cultural Revolution and Dengist reform).

Table 3.3

Date of Joining the People's Liberation Army

| Years | 1989 | | 1994 | |
|---|---|---|---|---|
| | No. | Percent | No. | Percent |
| 1929–1937 | 16 | 7 | 2 | 1 |
| 1938–1944 | 62 | 28 | 3 | 2 |
| 1945–1949 | 98 | 44 | 79 | 44 |
| 1950–1954 | 27 | 12 | 33 | 18 |
| 1955–1965 | 21 | 9 | 57 | 32 |
| 1966–1980 | 0 | 0 | 3 | 2 |
| Unknown | 0 | 0 | 2 | 1 |
| Total | 224 | 100 | 179 | 100 |

SOURCE: Mulvenon PLA database.

to their "elder superiors."[13] The key formative experiences of this generation were the Korean War, the Sino-Soviet alliance, the Sino-Indian Border War and the 1979 Sino-Vietnam border conflict. Swaine cautions that these officers should not be seen as "military professionals" (he prefers the term "military politicians," which emphasizes their involvement in elite politics) despite their participation in China's most "modern wars." He would agree, however, that these officers are more oriented toward modernization and professionalism than were their predecessors, highlighting again the theoretical distinction between "professionalism" as a static term and "degree of professionalism" as a fluid spectrum.

The primary weakness of the generational paradigm for our purposes, however, derives from the fact that the vast majority of our two data pools (and thus the entire high-level military leadership of the current PLA) are members of the second generation.[14] This would suggest that the explanatory power of the paradigm (as currently constituted) is greatest when applied to long-term historical surveys and is less useful for contemporary cohort analysis. For our

---

[13]Swaine, "Generations," p. 4.

[14]The fact that the second generation makes up the bulk of the military leadership should not come as a surprise but does suggest an interesting dynamic in Chinese military politics between elders with informal power but no formal power, second-generation officers with both formal and informal power, and third-generation leaders with neither formal nor informal power.

purposes, therefore, it is necessary to develop "intergenerational" variables for analysis, such as those found commonly in civilian co-hort analysis. In the case of the PLA, these would include place of birth, war participation, corps/group army affiliation, and profes-sional military education. The remainder of the chapter will be de-voted to examining each of these variables, assessing their individual usefulness in evaluating the officer cohort as well as their theoretical usefulness in supplementing the generational thesis.

## PLACE OF BIRTH

Place of birth is an important cohort variable because it is a potential source of extramilitary affiliation. Table 3.4 shows the distribution by birthplace of China's military leaders. The officers are divided by provincial origin, but in some cases subprovincial data were also ex-amined.

Most striking is that nearly one-quarter of the 1989 and 1994 officer corps was born in Shandong Province (24 percent). Additionally, large numbers of officers come from northern and eastern provinces such as Hebei (12 percent), Jiangsu (10 percent), and Liaoning (10 percent). This geographic distribution differs significantly from that of the early revolutionary era when the army's ranks were largely staffed by officers from Hubei, Jiangxi, and Hunan provinces.[15] Li and White call the Shandong result a "mystery" and wonder whether the phenomenon can be linked to Shandong guerrilla activity in the 1940s or regional favoritism in elite promotion. Recent interviews with PLA officers, however, offer a different explanation. They insist that the high incidence of Shandongese officers in the Chinese military is explained by Shandong's long and respected history of producing military leaders, which deeply influenced the career choices of province youth.[16] Additionally, they point out that

---

[15]Hubei, Jiangxi, and Hunan provinces contained many of the CCP's early "base areas" and were therefore logical sources of PLA personnel. See Whitson, *The Chinese High Command*, pp. 26–57.

[16]James Mulvenon, Personal interviews in Beijing, March 1997.

Table 3.4

PLA Officer Distribution of Birthplace, by Province

| Province | (% of Total Population) | 1989 No. | 1989 Percent | 1994 No. | 1994 Percent |
|---|---|---|---|---|---|
| **East** | | | | | |
| Shandong | (7.3) | 60 | 27 | 43 | 24 |
| Jiangsu | (5.9) | 22 | 10 | 17 | 10 |
| Anhui | (5.0) | 11 | 5 | 6 | 3 |
| Zhejiang | (3.6) | 4 | 2 | 4 | 2 |
| Shanghai | (1.1) | 2 | 1 | 1 | 5 |
| Jiangxi | (3.3) | 1 | 0 | 1 | 5 |
| Fujian | (2.7) | 0 | 0 | 1 | 5 |
| Subtotal | (28.9) | 100 | 45 | 73 | 40.5 |
| **North** | | | | | |
| Hebei | (5.4) | 46 | 21 | 21 | 12 |
| Shanxi | (2.5) | 10 | 4 | 9 | 5 |
| Tianjin | (0.8) | 3 | 1 | 1 | 5 |
| Beijing | (0.9) | 2 | 1 | 4 | 2 |
| Neimenggu | (1.9) | 1 | 0 | 2 | 1 |
| Subtotal | (11.5) | 62 | 28 | 37 | 20.5 |
| **Central/South** | | | | | |
| Hunan | (5.3) | 2 | 1 | 9 | 5 |
| Hubei | (4.8) | 7 | 3 | 7 | 4 |
| Henan | (7.6) | 12 | 5 | 6 | 3 |
| Guangdong | (5.6) | 1 | 0 | 3 | 2 |
| Guangxi | (3.7) | 0 | 0 | 0 | 0 |
| Subtotal | (27) | 22 | 10 | 25 | 14 |
| **Northeast** | | | | | |
| Liaoning | (3.4) | 13 | 6 | 17 | 10 |
| Jilin | (2.2) | 8 | 4 | 9 | 5 |
| Heilongjiang | (3.1) | 7 | 3 | 10 | 5.5 |
| Subtotal | (8.7) | 28 | 13 | 36 | 20.5 |
| **Southwest** | | | | | |
| Sichuan | (9.4) | 6 | 3 | 3 | 2 |
| Guizhou | (2.9) | 1 | 0 | 1 | 5 |
| Yunnan | (3.3) | 0 | 0 | 0 | 0 |
| Xizang | (0.2) | 0 | 0 | 0 | 0 |
| Subtotal | (15.8) | 7 | 3 | 4 | 2.5 |
| **Northwest** | | | | | |
| Shaanxi | (2.9) | 2 | 1 | 4 | 2 |
| Xinjiang | (1.3) | 2 | 1 | 0 | 0 |
| Gansu | (2.0) | 0 | 0 | 0 | 0 |
| Qinghai | (0.4) | 0 | 0 | 0 | 0 |
| Ningxia | (0.4) | 1 | 0 | 0 | 0 |
| Subtotal | (7) | 5 | 2 | 3 | 2 |
| Unknown | (0) | | | 0 | 0 |
| Total | (100) | 224 | 100 | 179 | 100 |

SOURCE: Mulvenon PLA database.

Shandongese recruits possess some valuable martial characteristics, such as above-average height and strength. These lines of analysis have been confirmed by one knowledgeable Western analyst, who asserts that these cultural and physical factors led the military leadership to aggressively recruit from Shandong province at the expense of other areas.[17]

Cultural explanations, however, may not be strong enough to explain all the irregularities in the data. For instance, Sichuan Province, whose population is the largest in the country (9.4 percent), accounts for only 2 percent of the 1994 officer corps. Additionally, Guangdong Province, which also had a cultural tradition of producing political and military elites during the Qing Dynasty, accounts for only 2 percent of the total. These results, along with the high rates of Shandongese mentioned earlier, cannot easily be dismissed as coincidences or outliers. At the same time, reports of a "Shandong faction" in the PLA must be viewed with appropriate caution and skepticism.[18]

The strongest explanation might be found in the geographic pattern of national reunification after 1945. Since most of the current leadership was recruited during or after the War of Liberation (see Table 2.3), there appears to be a relationship between geographic consolidation and distribution of recruits. For example, the regions that were liberated last (southwest, northwest, central-south) make up the three lowest percentages (18.5 percent total) of the officer corps' birthplaces, despite the fact that these provinces contain a large percentage of the overall national population (49.8 percent). In contrast, those regions that were liberated earlier (north, northeast, east) make up a much higher percentage of the officer corps (81.5 percent) than the overall population distribution (49.1 percent) would suggest. Of all the theories offered to explain the skewed distribution of birthplaces, this "consolidation" thesis seems to have the greatest

---

[17]Ibid.

[18]Tseng Hui-yen, "China's Military Power Gradually Falls Into Hands of 'Shandong Faction'; Officers Seek Promotion by Advocating Military Expansion Strategy," *Lien Ho Pao*, 21 October 1994, p. 10, in FBIS, 21 October 1994, pp. 34–35. The Hong Kong and Taiwan–based advocates of this theory have made the unwise logical leap from statistical correlation to causation, basing their conspiracy theories on the maldistribution of common geographic origin.

explanatory power.  At the same time, one knowledgeable U.S. military attaché believes that, as the second generation retires in favor of an entirely post-liberation generation, this distribution will probably become more uniform.[19]  If it does not, then an entirely new explanation must be sought.

As a supplement to the discussion of birth origins, two additional points of theoretical interest appeared in the course of this research.  The first deals with the issue of "native officers," or officers who serve in their native regions.  According to current regulations, no more than one-third of recruits in a military region can be native to the provinces of that region.  This percentage is even lower for units serving in politically sensitive areas such as Tibet, where only 10 percent of recruits can be locally born.  Generally, conscripts spend their entire tour in one military region.  Officers stay in a single military region until they reach the level of senior colonel, whereupon they can be moved horizontally as well as vertically through the levels of the military system.  This policy, which is similar in content to the imperial policy of the "law of avoidance," was designed to prevent the rise of regional power bases.[20]  The current governing principle is colloquially known by the Chinese aphorism *wu hu si hai*, "five lakes and four seas," meaning that military officers must be drawn from all corners of the country.[21]  To determine whether *wu hu si hai* is in effect for officers at the level of the military region, we compared the regional postings of officers in the data sets were compared to their birthplace.[22]

As seen in Table 3.5, the regional distribution of native officers suggests that there is not a universal application of a *wu hu si hai* in mili-

---

[19]James Mulvenon, Personal interviews in Beijing, March 1997.

[20]The law of avoidance refers to a system during the Imperial period of Chinese history that deliberately prevented government officials from serving in their home area for fear that they would develop strong local power bases.

[21]For a Hong Kong-based analysis of this principle, see Willy Wo-Lap Lam, "PAP Undertakes Thorough 'Changing of the Guards,'" *South China Morning Post*, 24 July 1996, p. 10.

[22]It must be noted, however, that this type of law of avoidance policy is not a defining characteristic of professionalism in a military.  At the same time, implementing a law of avoidance may smooth the transition to a professionalized PLA by eliminating obvious bases for factionalism.

Table 3.5

Portion of Military Region Leaders Born in
the Same Region

| | 1989 | | | 1994 | | |
| | | Born in Same MR | | | Born in Same MR | |
| | Total No. | No. | (%) | Total No. | No. | (%) |
|---|---|---|---|---|---|---|
| Region | | | | | | |
| Beijing | 13 | 5 | (38) | 11 | 4 | (36) |
| Jinan | 15 | 5 | (33) | 10 | 7 | (70) |
| Nanjing | 19 | 4 | (21) | 12 | 2 | (17) |
| Lanzhou | 21 | 2 | (10) | 15 | 1 | (7) |
| Chengdu | 14 | 1 | (7) | 12 | 1 | (8) |
| Shenyang | 18 | 1 | (6) | 13 | 0 | (0) |
| Guangzhou | 19 | 0 | (0) | 12 | 3 | (25) |
| Total | 119 | 18 | (15)[a] | 85 | 18 | (21)[a] |

SOURCE: Mulvenon PLA database.

[a]Weighted percentage.

tary appointments.[23] As in 1989, the 1994 pool shows that more than
one-third of the officers in the Beijing Military Region (MR) (36
percent), one-quarter of the officers in the Guangdong MR, and
slightly less than one-fifth of the officers in the Nanjing MR (17 percent) were native born. More striking is the fact that 70 percent of
the Jinan MR in 1994 were native officers, up from 33 percent in 1989,
even though officers from Henan and Shandong provinces (which
make up the Jinan Military Region) only account for 27 percent of the
officer corps. Thus, there is significant native representation in four
of seven military regions. Li and White dismissed similarly striking
results in their data (such as the 38 percent native complement in the
Beijing MR and the 33 percent complement in the Jinan MR), prefer-
ring to point to the 15 percent average across all military regions in
1989 as proof of a law of avoidance. Even if we accept this aggregation of the data, the total percentage of native officers in the seven
military regions still rose from 15 percent in 1989 to 21 percent in
1994.[24]

---

[23]Xu Shiyou, for instance, served as commander of the Nanjing Military Region for
more than 28 years.

[24]Of course, if the Jinan MR is taken out of the 1994 set, the weighted percentage of
native officers drops to 14.68, slightly less than the 1988 total.

This aggregation might be misleading, however, since it assumes that all military regions are of equal importance. For instance, it is logical to assume that the Lanzhou and Chengdu MRs might have low levels of native officers out of concern for the potential unreliability of minority officers during suppression of groups in Tibet or Xinjiang. If this is a calculated policy, however, it is also logical to question the high rates of native officers in two of the most important regions for China's internal security: Guangzhou MR and Beijing MR. The Guangzhou MR is vital to the stability of the reversion of Hong Kong in 1997, and the Beijing MR is central to the political stability of China's capital.[25] Indeed, it was reported that the local ties of certain officers during the Tiananmen crisis in 1989 led to some breakdowns in discipline and forced the central leadership to bring in units from outside the Beijing MR.[26] Yet the 1994 data suggest that no lesson has been drawn, as the percentage of native officers in the Beijing MR fell only slightly from 38 percent to 36 percent. This leads to one of two conclusions: (1) There is no policy of *wu hu si hai* for military region–level officers, or (2) the percentages above are within accepted limits in the eyes of the military and political leadership. To get a clearer picture, it would be necessary to analyze small groups of promotions or perhaps even the total pool of potential promotions at any given decisionmaking point. However, both of these empirical tasks would be extremely difficult or impossible with the current data.

At the same time, this significant native representation (along with the disproportionately coastal distribution of the officer corps seen in Table 3.4) does suggest some implications for potential PLA "regionalism," another popular subject in Western analysis.[27] One of the oft-mentioned sources of potential regionalism is the growing

---

[25]High native representation in the Guangzhou MR might help ameliorate the significant communication obstacles faced between speakers of Mandarin and Cantonese.

[26]See Timothy Brooks, *Quelling the People: The Military Suppression of the Beijing Democracy Movement*, New York: Oxford University Press, 1992.

[27]Of course, the vast majority of China's population lives in the eastern half of the country, and thus there might be a correlation between population and distribution of officers. On the other hand, the data seem to refute Beijing's propaganda about regional inclusion and representation in personnel decisions. For regionalism, see June Teufel Dreyer, *Regionalism in the People's Liberation Army*, in Richard Yang, ed., *CAPS Papers No. 9*, Chinese Council of Advanced Policy Studies, May 1995.

economic relationships between military units and local government officials. To replace expenditure funds, the CMC and the State Council have compelled local governments to increase their "contribution" to the support of military units in their geographic purview. One unexpected consequence of this decision was that local commanders and party officials began to join forces on both legal and illegal economic development projects. It is possible that these type of relationships would be further facilitated by the arrival of native officers, who, for both linguistic and fraternal reasons, would have an advantage over "outsiders."[28] Indeed, since the bulk of prosperity under the reforms is being generated in the coastal provinces, officers of coastal origin seem to be in a prime position to develop these types of linkages with local governments. But there is no evidence that this independent (and often corrupt) behavior has any political or military dimensions; it is therefore a less worrisome form of regionalism than the often-used terms "warlordism" or "independent kingdoms" would suggest.

## WAR PARTICIPATION

War participation is an important cohort variable because the bonds forged during combat are some of the most important in an officer's career and among the strongest sources of corporatism. Additionally, the type, outcome, and technological level of the combat are important influences upon the values and beliefs of the group, and affect its later attitudes toward specific policies.[29] Table 3.6 shows the distribution of war participation among the 1989 and 1994 officer corps. The numbers reveal a profound shift in experience. In particular, the number of officers who had fought in the Anti-Japanese War (1938–1944) fell from 46 percent to only 2.5 percent, while the number officers whose first war was the Civil War (1945–1949) or the Korean War (1950–1953) rose from 30 percent to 51 percent. This confirms the observations of Michael Swaine, Paul Godwin, and others that the Chinese officer corps has undergone a tectonic shift of experience from the Revolutionary War period to the

---

[28]This type of economic "regionalism" is distinct in character from political regionalism, although for bureaucratic reasons the two are not mutually exclusive.

[29]An analogy would be the experience of the U.S. military in Vietnam, after which many of the military establishment's core beliefs were questioned.

Table 3.6

**Distribution of War Participation**

|  | 1989 | | 1994 | |
|---|---|---|---|---|
|  | No. | Percent | No. | Percent |
| 1st War |  |  |  |  |
| RW | 9 | 4 | 2 | 1 |
| AW | 98 | 44 | 3 | 2 |
| LW | 56 | 25 | 81 | 45 |
| KW | 11 | 5 | 11 | 6 |
| Other war | 2 | 1 | 1 | .5 |
| Specific experience |  |  |  |  |
| RW, AW, LW, KW | 6 | 2 | 1 | .5 |
| RW, AW, LW | 2 | 1 | 1 | .5 |
| RW, AW, KW | 1 | 0 | 0 | 0 |
| RW, AW | 0 | 0 | 0 | 0 |
| AW | 1 | 0 | 0 | 0 |
| AW, LW | 52 | 23 | 1 | .5 |
| AW, KW | 7 | 3 | 0 | 0 |
| AW, LW, KW | 38 | 17 | 2 | 1 |
| LW | 25 | 11 | 35 | 20 |
| LW, KW | 31 | 14 | 46 | 25 |
| KW | 11 | 5 | 11 | 6 |
| OW | 2 | 1 | 1 | .5 |
| No war experience | 48 | 21 | 81 | 46 |
| Total | 224 | 100 | 179 | 100 |

SOURCE: Mulvenon PLA database.
NOTES:  RW = Revolutionary War (1927–1937), AW = Anti-Japanese War (1937–1945), LW = War of Liberation (1945–1949), KW = Korean War (1950–1953).

Civil War and Korean War.[30] The importance of this change cannot be overstated since it symbolizes the very evolution of the PLA from a guerrilla army to a more modern fighting force.  In particular, the officers who fought in Korea, most prominently Peng Dehuai and Liu Bocheng, were left with a profound appreciation of the needs of modern technological warfare and were the most vocal proponents of the PLA's modernization (with Soviet help) in the 1950s.  While the personal woes of Marshal Peng and the vicissitudes of the Cultural

---

[30]The author would like to thank Paul Godwin for his comments on this subject.

Revolution tabled this agenda for more than twenty years, the veterans of this period now make up the bulk of China's military leadership. Their imprint can be seen on many of the PLA's modernization and doctrine reforms in the 1980s, such as "limited war under high-tech conditions."

Equally striking is the increase in the number of officers (from 21 percent to 46 percent) who have no combat experience at all. The average age of these 76 officers in 1994 was 58.13, which is significantly less than the corps average of 61.45. This disparity suggests a number of persuasive reasons why these younger officers should have no battle experience. First and foremost, a 58-year old officer would have been 14 when the Korean War broke out and thus would have missed his prime opportunity for combat experience. Granted, China has engaged in military conflict since then (e.g., Taiwan in 1954–1955 and 1958, India in 1962, Vietnam in 1979), but these conflicts were relatively small in scale and afforded less opportunity for battlefield experience. Second, the technical modernization of the PLA has resulted in the promotion of a significant number of technocrats, who, by virtue of their specialization, have never seen combat. Third, there are unfortunately some exceptions to the rule of merit-based promotion, such as Wang Ruilin, whose service as Deng Xiaoping's secretary has propelled him through the ranks.[31]

## PROFESSIONAL MILITARY EDUCATION

As discussed in Chapter One, PME attendance exerts important, if sometimes contradictory, influences upon the cohort. On one hand, a common school tie or class ring makes PME graduates a more cohesive pressure group and therefore more politically powerful, which is contrary to the Huntingtonian notions of professionalism apoliticism. In the Chinese case, moreover, high-level PLA institutions like the NDU provide an opportunity for cross-branch, cross-regional interaction that the vertical hierarchy and party committee system

---

[31]One popular explanation for Wang Ruilin's recent promotion to the CMC is that he has transferred his political loyalty to Jiang Zemin and is helping the latter weed out the more recalcitrant "Dengist" elements and/or build a bridge between Jiang Zemin and Deng loyalists. If this true, it merely highlights the nonprofessional nature of his military career.

had previously prevented. On the other hand, these "joint" interactions are necessary if the PLA hopes to increase the level of jointness in its combat operations. Furthermore, some scholars argue that PME-based networks can also be beneficial to the functioning of a professional military since PME attendance creates a sense of professional corporatism and permits the construction of a "system of sponsorship," by which high-ranking officers are able to influence the careers of promising young officers by requesting their assignment to their own staffs or recommending them for appropriate posts. According to this hypothesis, the contacts formed among these groups of peers and superiors eventually become the dominant influence on an officer's career.

In the Chinese case, one of the most popular "system of sponsorship" arguments centers on CMC Vice-Chair General Zhang Zhen, who was president of the NDU from 1985 to 1992 before his ascension to the CMC with the portfolio for personnel decisions. According to this hypothesis, General Zhang has tended to promote officers with whom he had developed relationships at NDU, thus creating an NDU clique centered on himself in the upper ranks of the officer corps. Over the years, this interpretation has reached the level of myth, although no empirical evidence has ever been offered to validate it.

Our data show that no members of the 1989 pool had attended the NDU during Zhang's tenure. In 1994, after Zhang became CMC Vice-Chairman for personnel decisions, the number was only 2 officers (less than 1 percent).[32] There are three possible explanations for this low result. First, the data could be incomplete, although NDU attendance was prominently touted in the Chinese sources. Second, the NDU students who attended during Zhang Zhen's tenure might not have reached MR-level postings, and therefore were absent from the pool. Third, the hypothesis is simply invalid. Overall, we are inclined to believe a combination of all three reasons, though incomplete data probably deserve the lion's share of the blame. It is highly implausible that so small a percentage of the officers in the pools would have attended the NDU, particularly given the fact that no Chinese officer can achieve flag rank without attending the programs.

---

[32]Mulvenon PLA database.

Even if we could correct the data, however, it would be misguided to assume that NDU graduates automatically owe any allegiance to Zhang Zhen or any other school official. No commandant of a military school, Chinese or otherwise, has the time to develop deep relationships with the hundreds of students who pass through the gates, except perhaps a few exceptional students brought to his attention. Also, some analysts believe that education at NDU occurs too late in many officers' careers to significantly alter their situation, which directly contradicts the research of Trout and others.[33] As one military attache pointed out, officers are more likely to be influenced by the experience of common service in their original units, and those bonded relationships are believed to be far stronger than a short stint at a high-level military institution.[34] Dreyer points out that this dynamic is well-established in the sociological literature, which concludes that "most individuals develop a core of political loyalties and attachments which are generally resistant to change in later life," except under severe pressure.[35]

Another major criticism of the military education thesis is more statistical in origin and centers on the concept of "sampling on the dependent variable," in this case incidence of attendance. Since more than 50 percent of the officers in the two pools attended a military school, it is not particularly meaningful to use school attendance to distinguish among the cohort members unless more specific details, such as class year, are known. If these data were available, they could prove very significant since the most potentially interesting component of military education is intraclass relations between students, which then might become the lattice upon which future networks are based.

---

[33]David Moore and B. Thomas Trout, "Military Advancement: The Visibility Theory of Promotion," *American Political Science Review,* 72, 1978, pp. 452–468.

[34]James Mulvenon, Personal interviews in Beijing, March 1997. Paul Godwin dissents somewhat from this view, arguing that attendance at high-level military schools inculcates a sense of elite status among the chosen officers. He suggests that this elite status does not eradicate earlier bonding experiences but creates a complementary bond that crosses service boundaries.

[35]Dreyer, "The New Officer Corps," p. 330.

## CORPS AFFILIATIONS

Michael Swaine and others persuasively argue that the corps/group army is now a more valuable level of analysis than field armies, because virtually all of the officers in the Korean War generation began their military careers as junior officers within a specific field army's corps during the late forties.[36]  These individuals continued to serve in the same corps during the Korean War, mainly at the regimental and divisional levels, and usually continued to rise through the ranks of their original units for decades thereafter.  Affiliation with this "home" corps was broken only by dissolution of the corps itself following the Korean War, short stints at a military school, or a staff post in a regional MR  and did not officially end until the officer retired or had risen into the ranks of the national military leadership.[37]  Thus for the majority of these officers, association with an individual corps took precedence over the initial identification with an entire field army, and this association provided the foundation for the factional network necessary to upward mobility in the PLA.[38]  This is not to say that the field army had no influence upon these young officers' careers, since common field-army affiliation served as the basis for the establishment of upward links to powerful patrons and the networks for crisis mobilization as late as the early 1990s.[39]  The death of many, if not nearly all, of the military elders in recent years, however, may have begun to break this historical and political connection, eventually transforming corps and group armies into fully discrete units of analysis.

The corps affiliations for officers in the two data pools are summarized in Table 3.7.

At first glance, the remarkable uniformity of the distribution suggests that no one corps or corps commander has exerted an overwhelming influence over promotion patterns in the PLA.  It could be argued that this outcome verifies the commonly held assumption that the Chinese military and political leadership consciously tries to achieve

---

[36]Swaine, *Military and Political Succession*, p. 25.

[37]For a discussion of "uprooted cadres," see Ibid., p. 25.

[38]Ibid., p. 142.

[39]Ibid., p. 25.

Table 3.7

**Corps/Group-Army (GA) Affiliations**

| Corps/GA | 1989 Pool | | 1994 Pool | |
|---|---|---|---|---|
| | Confirmed | Tentative | Confirmed | Tentative |
| 1st | 2 | 3 | 2 | 2 |
| 2nd–10th | 0 | 0 | 0 | 0 |
| 11th | 1 | 0 | 1 | 0 |
| 12th | 2 | 1 | 2 | 2 |
| 13th | 0 | 0 | 0 | 0 |
| 14th | 0 | 1 | 0 | 2 |
| 15th | 2 | 1 | 0 | 1 |
| 16th | 1 | 0 | 3 | 1 |
| 17th | 0 | 0 | 0 | 0 |
| 18th | 0 | 1 | 1 | 1 |
| 20th | 0 | 1 | 0 | 1 |
| 21st | 2 | 1 | 3 | 1 |
| 22nd | 0 | 0 | 0 | 2 |
| 23rd | 2 | 3 | 1 | 2 |
| 24th | 1 | 0 | 1 | 2 |
| 25th | 1 | 0 | 0 | 0 |
| 26th | 3 | 3 | 1 | 3 |
| 27th | 2 | 0 | 1 | 0 |
| 28th | 1 | 1 | 2 | 0 |
| 29th, 30th | 0 | 0 | 0 | 0 |
| 31st | 3 | 0 | 3 | 2 |
| 32nd–37th | 0 | 0 | 0 | 0 |
| 38th | 2 | 1 | 3 | 1 |
| 39th | 3 | 1 | 4 | 3 |
| 40th | 3 | 1 | 1 | 3 |
| 41st | 4 | 4 | 3 | 5 |
| 42nd | 2 | 0 | 4 | 0 |
| 43rd | 2 | 4 | 2 | 3 |
| 46th | 1 | 0 | 1 | 2 |
| 47th | 1 | 2 | 3 | 2 |
| 48th, 49th | 0 | 0 | 0 | 0 |
| 50th | 1 | 0 | 1 | 0 |
| 54th | 3 | 3 | 3 | 3 |
| 55th | 0 | 4 | 0 | 3 |
| 56th–59th | 0 | 0 | 0 | 0 |
| 60th | 2 | 1 | 0 | 1 |
| 61st, 62nd | 0 | 0 | 0 | 0 |
| 63rd | 2 | 1 | 2 | 4 |
| 64th | 1 | 1 | 3 | 2 |
| 65th | 0 | 1 | 0 | 4 |
| 66th | 0 | 0 | 0 | 3 |
| 67th | 0 | 2 | 1 | 3 |
| 68th | 0 | 1 | 0 | 2 |
| Total | 50/224 | 43/224 | 53/179 | 66/179 |

SOURCE: Mulvenon PLA database.

a sense of balance in promotions, preventing any large group of officers from a particular background from congregating at the top of
the military hierarchy. Yet the evidence also suggests that the members of some corps are more successful than others. The best explanation for the distribution of corps affiliations seems to derive from
the evolutionary history of the corps system, in particular the variable pattern of corps survival over the years. In the aggregate, 80
percent of the identifiable corps affiliations in the 1989 pool and 85
percent of the affiliations in the 1994 pool involve corps that have
survived the multitude of consolidations and deactivations that oc-
curred after Liberation, the Korean War, and the 1985 reduction in
force (RIF). While a minority of officers have been promoted despite
the deactivation or consolidation of their home corps, the vast
majority of the sample rose through the ranks of corps that still exist
in group army form today. This empirically verifies the general fate
of "uprooted cadres" described in Michael Swaine's 1992 study on
the Chinese military leadership, in which he argues that uprooted
cadres are "exceptions" in the top PLA leadership since the loss of
one's factional base has nearly always prevented promotion to the
top ranks.[40]

However, the fact remains that the number of identifiable
corps/group-army affiliations makes up a very low percentage of the
overall sample. Only 28 percent of corps affiliations in the 1988
group could be identified, dropping slightly to 24 percent in the 1994
pool. This outcome can be traced to three factors. First, data on the
specific unit history of officers are woefully inadequate. Often an officer's career is described in vague terms, identifying his successive
positions but not the corresponding units. Second, the corps themselves have not remained static over time. As Swaine's fold-out extension of Whitson's classic chart shows, many corps were consoli-
dated into group armies or eliminated completely between 1953 and
1985.[41] These changes served to disrupt or destroy some of the potential factional lattices provided by corps and left many military officers with no clear sense of subinstitutional identity.

---

[40]Ibid.

[41]Ibid., p. 243.

Third, the corps argument appears to be appropriate only for ground force officers and to a slightly lesser extent for artillerymen and tankers.[42] Naval, air force, and 2nd Artillery officers followed very different career paths, which were almost completely separate from the corps system. For example, in the 1989 and 1994 data pools, there were 25 and 22 naval officers, respectively. Ellis Melvin argues persuasively that these officers followed a very different path than did their ground-force counterparts:

> The PLA Naval officers in this age group do not have ground force service time but went straight into the navy. Their service time was spent on a naval base. Some came up through the warship divisions and squadrons and could be sorted by type of warship. Naval air force officers would come up through flights, squadrons, regiments, divisions, and fleet naval air forces.[43]

For the 48 and 22 air force officers in the two pools, the circumstances are very much the same:

> The PLA Air Force officers in this age group do not have ground force service time but went straight into the Air Force. An Air Force officer would have service time on an air army or Air Force command post. Some came up through the flights, squadrons, regiments, and divisions. In the airborne units, the officers would come up through the companies, battalions, regiments, and divisions to the airborne army posts.[44]

Finally, the Second Artillery officers followed a third, unique non-ground-force path:

> The PLA 2nd Artillery (Strategic Rocket Forces [SRF]) officers in this age group do not have ground force service time and generally went straight into the Strategic Missile Force or transferred in from

---

[42]Armored and artillery officers (e.g., Shi Baoyuan) should be considered part of the group-army structure and, in those cases where they are at region or district, within that structure, since they will be promoted within the group-army organizations. There are still some units directly under the Military Branches Department of the Headquarters of the General Staff, but their numbers are declining as time goes by. I am grateful to Ellis Melvin for this observation.

[43]James Mulvenon, Personal correspondence, 1 April 1997.

[44]Ibid.

COSTIND units that deal with missile testing. These officers worked their way up through batteries, battalions, brigades, and bases, the latter of which is equivalent to an army. The general officers of the 2nd Artillery follow a path from the colleges to the PLA to the various 2nd Artillery schools and up through the units, with the majority being in the command career fields. The political officers who have experience in handling troop affairs for remote units sometimes transfer to political posts in the SRF. In the past, the officers came from artillery, public security forces, and the Railway Corps, primarily. The artillery officers were used in the early launch sites. Public Security Force officers were administrative holdovers from the old Public Security Forces Headquarters when the building was transferred to the 2nd Artillery HQ in July 1966 and became the political officers of the Second Artillery. The Railway Corps officers had experience in tunnel digging and served as engineers in building missile sites.[45]

For non-ground-force officers, therefore, the corps/group army is a less appropriate level of analysis and must be replaced with service-specific career markers.

An additional problem is presented by political officers. When looking at the data, one is sorely tempted to subdivide the officer corps between political and nonpolitical officers since the career paths and promotion standards of the former are not the same as the latter. Ellis Melvin argues that, organizationally, political officers in all of the services should be considered career political officers since their advancements follow different paths than their commander counterparts.[46] Table 3.8 shows the number of PLA career political work officers in the 1989 and 1994 data pools. The biographical evidence strongly suggests that these political officers were promoted independently from command officers, according to the dictates of the

---

[45]Ibid.

[46]Specifically, the political work system within the PLA is divided into three vertical structures (the political commissar system, the party committee system and the discipline inspection system) down to the regimental level, below which the arrangement is ad hoc. Of these three, the most relevant for this report is the political commissar system since most of the political officers in the two data pools rose through the political commissar system. For an excellent discussion of the political work system in the PLA, see Shambaugh, "The Soldier and the State."

Table 3.8

Number of Career Political Work Officers in the PLA

| Service | 1989 | 1994 |
|---------|------|------|
| PLA ground forces | 43 | 54 |
| PLA Navy | 10 | 7 |
| PLA Air Force | 14 | 10 |

SOURCE: Mulvenon PLA database.

General Political Department.[47] Thus, although political officers technically served in corps, these units do not define their institutional identity and should not be used to analyze their background.

Given these findings, what can we conclude about the cohort dynamic in the current officer corps? There are disparities in the geographical distribution of officers, but this distribution can be best explained by the timing of liberation and consolidation in the late 1940s and early 1950s. In terms of war experience, the starting point for the current officer corps is the Korean War and the "modern" wars that followed, which presents a powerful constituency for technological modernization and post–"People's War" doctrines. In terms of NDU attendance, less than 1 percent of the 1994 pool had attended the NDU during Zhang Zhen's tenure, which either disproves the commonly held assumption that Zhang Zhen, in his current CMC position with the portfolio for promotions, has consciously promoted former NDU students that he meet during his presidency or exposes a serious gap in the data. Finally, the corps affiliations of officers in both pools are remarkably uniform, which suggests that no one corps or corps commander has exerted an overwhelming influence over promotion patterns in the PLA. Yet the evidence also suggests that the members of some corps are more successful than members of other corps. While a minority of officers have survived despite the deactivation or consolidation of their home corps, the vast majority of the sample rose through the ranks of corps that still exist in group-army form today.

---

[47]Additionally, these promotions were and continue to be based on different criteria than those for commanders, although the standards are changing as the definition of political work in the PLA evolves. Discussion of the changes in the definition of political work can be found in the concluding chapter.

Overall, therefore, the corps/group-army variable has the greatest future potential of all the cohort variables examined in this study, particularly with regard to the connection between cohorts and networks. With the limited current data, it is still possible to identify probable connections between officers based on their early service in specific corps. With better data, these linkages might be solidified, providing a powerful tool for analyzing internal Chinese military behavior.

# Chapter Four

# CONCLUSIONS

## ASSESSMENT OF CHINESE PROFESSIONALIZATION

The data suggest that the PLA officers corps between 1989 and 1994 has become younger, better educated, more functionally specialized and more subject to an institutionalized retirement norm. Changes between 1989 and 1994 also suggest that these trends will continue in the future, as older (and often less professional) officers are retired in favor of their more professional counterparts. Eventually, the average age and education level of the PLA may stabilize at a equilibrium that satisfies both the need for professional officers and the logistical requirements of the army's new promotion and retirement mechanisms. These trends are all positive developments for Chinese military professionalization, since they improve the army's expertise, rationalize career patterns within the leadership, and build the foundation for Huntington's concept of corporateness. On a comparative level, they confirm the technocratic transition occurring across the entire Chinese political scene, and provide further evidence of a fundamental elite transformation from revolutionary cadres to professional managers. However, it is difficult to make a definitive statement about the current level of professionalization in the PLA since the term includes far more dimensions than the data could address. Nonetheless, the areas under examination show significant and lasting improvement and provide a solid base for further professionalization.

The more problematic portion of the analysis concerns cohort analysis. Here, the ambiguous nature of the data makes hard conclusions difficult. Working from the assumption that Whitson's field-army thesis now has only marginal explanatory power because of the deaths of important elders, we examined a variety of alternative frameworks, the most promising of which appears to be generational analysis. As outlined above, however, generational analysis is less useful for the purposes of this report because the majority of the current military leadership is drawn from the Korean War generation, whose formative experiences included the Korean War, the Sino-Soviet alliance of the early 1950s, the 1962 Sino-Indian border war, the 1969 Sino-Soviet border clash and the 1979 Sino-Vietnamese border war. As a result, the analysis focused on a set of variables relevant to the current officer corps: place of birth, war experience, corps affiliation, and military education.

We conclude that the PLA has undergone a profound generational shift from the revolutionary generation to a new post-Liberation cohort. The starting point for the current officer corps is the Korean War and the "modern" wars that followed, making them a powerful constituency for technological modernization and post–People's War doctrines. In terms of NDU attendance, less than 1 percent of the 1994 pool had attended the NDU during Zhang Zhen's tenure, which must be a data error. Nonetheless, this report rejects the commonly held assumption that Zhang Zhen, in his current CMC position with the portfolio for promotions, has consciously promoted former NDU students that he meet during his presidency. Finally, the corps affiliations of officers in both pools are remarkably uniform, suggesting that no one corps or corps commander has exerted an overwhelming influence over promotion patterns in the PLA. Yet the evidence also suggests that the members of some corps are more successful than others. While a minority of officers have survived despite the deactivation or consolidation of their home corps, the vast majority of the sample rose through the ranks of corps that still exist in group-army form today. Overall, the corps/group-army variable has the greatest future potential of all the cohort variables examined in this study, particularly with regard to the connection between cohorts and networks. With better data, these linkages might be solidified, providing a powerful tool for analyzing internal Chinese military behavior.

These trends help clarify the dynamic of affiliations within the PLA, which appears to be changing in character from the Maoist and early reform era. It is unrealistic to expect any large, differentiated organization to be devoid of personalistic relationships altogether. It is more important to identify the dynamics and limits of internal behavior within an organization and assess the potential impact of these cleavages upon the effectiveness of that organization. These data also help to clarify the personalistic dynamics in the PLA, which appear to be changing in character from the earlier faction-ridden era. Overall, the PLA could be said to be developing more professional-type networks (similar in some respects to the U.S. military), latticed around traditional personal ties as well as professional military education, field performance, and other avenues of professionalization. These networks have a contradictory effect upon the military: They strengthen the PLA by creating associational groups and providing additional sources of information for the promotion process, yet they also weaken the PLA by creating new communication and power channels outside of the traditional chain of command. Some of these latter negative consequences, however, are mitigated by the fact that the professionalizing trends in the PLA ensure that there is a rising meritocratic "floor," permitting cohort-based affiliations to serve as a mechanism for differentiating among a cohort of largely professional and competent officers.

On balance, therefore, the evidence suggests that the PLA is becoming more professionalized in the conventional sense of the term. The true degree of PLA professionalization and the potential impact that the process will have upon the system is less than definitive, however, since it is also affected by interaction with economic, political, and social trends in the nation as a whole. Yet it is clear that the PLA is certainly more professional than both its prereform and midreform antecedents, and these trends show no signs of reversing. If anything, they are accelerating. Assuming that current trends persist, the PLA officer corps of the future will increasingly resemble Western militaries, though the needs of economic development and the nature of the Chinese political regime will continue to serve as structural constraints upon this process.

If we revisit the Huntingtonian model and assess the PLA's professionalization according to the criteria of corporateness, responsibility, and expertise, some possible findings emerge. Each of these concepts is difficult to measure in any definitive fashion, so the analysis presented must be treated as informed speculation.

Corporatism has always been a thorny concept for students of communist militaries, in part because ideology often created the same type of intensely strong bonds between soldiers as fostered under Huntingtonian professionalism. Historically, for instance, the PLA has never had a problem distinguishing itself as a corporate identity apart from laymen, though the divisiveness of the Cultural Revolution and Tiananmen sorely tested the army's organic unity. Despite Mao's dislike of rank and status in the military hierarchy, there was still a profound pride associated with military service and a lifelong bond among veterans. While this may simply be a semantic argument over the definition of corporatism, the data suggest a more compelling thesis: The military's specific type of corporatism is determined by its sources, and therefore professional corporatism is just one end of a widely varied spectrum. Thus, the source of PLA corporatism is changing from the previous ideology-based egalitarianism to reflect a more professional ethic, based on rank, hierarchy, and pride in craft. Whereas during the Mao era the criteria for advancement may have been ideological, current criteria are weighted more heavily toward competency and skill, though political reliability is still a prerequisite. Thus the reforms of the 1980s and 1990s (e.g., reissue of ranks, professional military education, competency-based promotions) combined with the pride and nationalism of China's status as a rising regional power cannot help but inculcate a growing professional esprit de corps in the PLA, though the degree will always be open to debate. At the same time, this stronger corporate identity, when combined with the CCP's insistence that the armed forces remain engaged in internal domestic affairs, leads to the potential for a praetorian PLA.

In terms of responsibility, the PLA seems to have a firm grasp on one-half of the equation. On the one hand, the fierce nationalism of the officer corps alluded to above, exhibited in their reputed pressures on civilian foreign policy, ensures that the PLA will never shirk from its duty to defend the motherland from foreign invasion or assert its "soft frontiers."[1]  On the other hand, the "client" of the PLA is still the CCP and not society or the state in an abstract sense—as seen in recent statements by Jiang Zemin that do not preclude the use of military force to crush internal dissent, despite the alleged displeasure of elements in the military leadership over Tiananmen. As long as the political direction of the army is controlled by the party (usually embodied in one dominant person), the PLA will remain governed by "subjective civilian control" (i.e., maximizing civilian power at the expense of military power) rather than Huntington's preferred "objective civilian control" (i.e., professionalization of the military). On this point, however, it is important to note that political study in the PLA no longer focuses on the specific philosophical tenets of particular ideology espoused by the CCP (e.g., Maoism) but instead concentrates on the more instrumental ethos of single-party control of the PLA by the CCP, embodied in Deng's work on army-building in the new period and Mao's dictum about the "party" always controlling the gun. This can be strongly contrasted with the definition of responsibility in more professionalized Western militaries, which focuses almost exclusively on the administration-neutral notion of "duty, honor, country."

Perhaps the closest fit between Huntington's theory and the current Chinese officer corps can be found in the PLA's increasing expertise, manifested in the high number of officers with college or military education and the institutionalized relationship between advanced education and promotion. This rise in technical and theoretical proficiency is essential to the logistical, tactical, and strategic operation of a modern, technologically advanced military organization. At the same time, expertise is a double-edged sword. Military proficiency can be used for nonprofessional purposes, and the professional military education system can be used for network building. This reflects Bengt Abrahamsson's belief that a professional army may simply be a more unified and capable intervening force in domestic politics.[2] On the other hand, if our definition of professionalism were so nar-

---

[1]"Soft frontiers" refers to spheres of hegemonic influence as opposed to "hard frontiers" like national boundaries. China counts most of continental Asia within its soft frontiers, which is used to legitimize behavior like the 1979 Sino-Vietnamese border conflict.

[2]This conclusion is extrapolated from the excellent and often overlooked work by Bengt Abrahamsson, *Military Professionalism and Political Power*.

row as to exclude school affiliation, then no military on the globe, including the U.S. military, could truly be considered professional in the purest sense. In the case of the PLA officer corps, however, the effect of a rising educational level upon its professional character outweighs the potentially negative consequences of school affiliation because the meritocratic floor of the officer corps is higher than in the past.

Thus, the discussion returns to relative versus absolute gains. In absolute terms, the PLA is not professional in the Huntington sense nor will it ever be, as long as there is a party committee system in the army and single-party rule by the CCP. In relative terms, however, the current PLA is much more professional in character than was the army of the Mao period or even the pre-1985 army. The continuing nonprofessional aspects of the PLA are a function less of the institution than of the ideological and political milieu in which the institution operates. As that milieu evolves, the definition of PLA professionalism will evolve. In a continued single-party rule context, the professionalism of the PLA will derive from its increasing military competence and capability. Under a more pluralist type of political system, the definition of PLA professionalism will have to be expanded to include an assessment of the military's political character, especially the degree of its apoliticism.

One serious mitigating factor that has not been discussed in this report, however, is the PLA's rapidly expanding participation in the economy, both legal and illegal. Ellis Joffe persuasively argues that this economic behavior presents one of the greatest threats to continued professionalization of the PLA.[3] He argues that economic entanglement could have adverse effects upon the PLA and its relations with the Party. Specifically, the widespread corruption spawned by these activities could erode the ethic of duty central to professionalism and demoralize the troops by diluting their pure military identity. It could degrade military readiness, especially if troops are diverted to economic pursuits at the expense of military tasks. It threatens to weaken the corporate bonds between officers because competition over economic opportunities could foster rivalries between localized units. It could disrupt specialized training, which

---

[3]See Ellis Joffe, "Party-Army Relations in China," pp. 311–312.

requires the continuous acquisition of knowledge and skills. It could destroy discipline because important access to economic connections might not reflect the chain of command, and officers involved in economic activities may circumvent orders that threaten profits. Finally, participation in commercialism might threaten Party control, particularly given the center's mandate that local military leaders rely increasingly on local civilian leaders for financial support.

Despite these real and potential consequences of economic participation, both the political and military leaderships have been reluctant to curtail it. The military leadership realizes that these enterprises often employ soldiers' dependents and demobilized soldiers, as well provide units with desperately needed funds for housing and messing, thus reducing the burden on the center to meet these needs. The political leadership, ever cognizant of the central role of the PLA in the post-Deng transition, is understandably reluctant to separate the PLA from its businesses unless it can replace the lost funds with increased budgetary allocation.[4] The evidence suggests that this is probably bureaucratically and fiscally impossible. For the forseeable future, therefore, the PLA is trapped in a Catch-22: It cannot fully professionalize without abandoning its economic enterprises nor can it sustain the current professionalization process without them. The only solution may be a regimen of drastic troop cuts (so that the current budget dollar goes farther), followed by a gradual weaning of the PLA from commercial activities.

## IMPLICATIONS FOR CIVIL-MILITARY RELATIONS

One major question deals with the PLA's relation to the state (in this case the CCP headed by Jiang Zemin) and how that civil-military dynamic might be affected by the professionalizing trends among the officer corps outlined above. The dynamics of civil-military relations in Leninist and non-Leninist systems have been widely discussed in other studies, so this report will only offer a few caveats.[5] Despite

---

[4]The Hong Kong press has published reports in which Jiang Zemin laments the situation, claiming that if he had an additional U.S.$5 billion, he would give it to the PLA in exchange for a clean break from the economy.

[5]For a good review of the Cold War–era literature, see Amos Perlmutter and William M. LeoGrande, "The Party in Uniform: Toward a Theory of Civil-Military Relations in

important changes in the Chinese system, there is little evidence to suggest that the PLA is withdrawing from politics.[6] At the highest level, Liu Huaqing's membership on the Politburo Standing Committee since 1992 suggests a continued or even enhanced role for the military in Chinese domestic politics, especially in the post-Deng environment. There is every reason to expect that Admiral Liu's place on the Standing Committee will be filled with another PLA officer (either Chi Haotian or Zhang Wannian) if Liu retires at the Fifteenth Party Congress in fall 1997. This institutionalized political role is bolstered by PLA participation in key foreign-policy related fora, especially those that focus on relations with Taiwan and the United States.[7] Additionally, the remaining PLA elders, such as Zhang Aiping, continue to wield substantial, albeit declining, influence. The final and perhaps most powerful influence exercised by the PLA does not involve any single individual or organization but is symbolized by the institution itself. In an atmosphere of uncertainty and transition, the perceived preferences of the PLA serve as policy constraints on the political leadership and Jiang Zemin. Whether the PLA acts as kingmaker or influences policy more passively as a veto actor, its post-Deng position suggests that Jiang and the civilian leadership cannot make a major decision in the elite politics or foreign policy realms without first assessing the preferences of the military.

This line of argument highlights the nexus between professionalism and civil-military relations. Accepting the premise that militaries in Leninist systems will never be fully divorced from the political arena, increased professionalism in the officer corps has the paradoxical

---

Communist Political Systems; *American Political Science Review*, and Dale Herspring and Ivan Volyges, *Civil-Military Relations in Communist Systems*, Boulder, CO: Westview Press, 1978. For more recent examinations of the subject, see Ellis Joffe, "Party-Army Relations in China," pp. 299–300; and David Shambaugh, "The Soldier and the State in China: The Political Work System in the People's Liberation Army." For an excellent literature review, see Jeremy Paltiel, "PLA Allegiance on Parade: Civil-Military Relations in Transition," *China Quarterly*, No. 143, September 1995, pp. 784–800.

[6]For an excellent review of the current political campaigns within the PLA and the intricacies of its various political systems, see Shambaugh, "The Soldier and the State in China."

[7]For a full examination of this subject, see Michael Swaine, *The Role of the Chinese Military in National Security Policymaking*, Santa Monica, CA: RAND, MR-782-OSD, 1996.

outcome of strengthening the military's advocacy of its political positions. According to Abrahamsson, professionalism leads to corporatism, which strengthens unity and in turn strengthens the ability of military leaders to present a united front to the political leadership.[8] Abrahamsson's counterintuitive assertion is a perverse twist on Huntington's link between professionalism and apoliticism. Given the importance of the PLA in the coming post-Deng succession, this potentially strengthened unity of the PLA could make it a very powerful determinant of Chinese domestic politics for the foreseeable future. Abrahamsson's concept also serves as the missing link in the operationalization of Joffe's notion of a "Party-Army with professional characteristics"[9] because it builds a bridge between the contradictions of professionalism and continued military involvement in politics.

## IMPLICATIONS FOR MILITARY MODERNIZATION AND OPERATIONAL EFFECTIVENESS

Another potential implication of increased professionalism in the Chinese officer corps is improved operational effectiveness. High-tech war, embodied in the conduct of the 1991 Persian Gulf War, demands a high level of base education (both civilian and military) as well as rigorous and realistic training. All evidence suggests that the current PLA officer corps has made great strides in these areas, particularly in comparison to its predecessors. Indeed, the exercises of March 1996 showed that the PLA has come a long way since the disastrous 1979 border conflict with the Vietnamese, where the flaws of Mao's People's War doctrine were tragically exposed to outsiders as well as to the Chinese military leadership. Since that time, the PLA has successfully (if somewhat haltingly) revamped its entire internal doctrine[10] and has begun the difficult task of operationalizing that doctrine through training at increasingly higher units of military or-

---

[8]Abrahamsson, *Military Professionalism.*

[9]Joffe, "Party-Army Relations in China," p. 300.

[10]See Li Nan, "The PLA's Evolving Warfighting Doctrine, Strategy, and Tactics, 1985–95: A Chinese Perspective," *China Quarterly*, No. 146, June 1996, pp. 443–463; and Paul H. B. Godwin, "From Continent to Periphery: PLA Doctrine, Strategy, and Capabilities Towards 2000," *China Quarterly*, No. 146, June 1996, pp. 464–487.

ganization.[11]  There is no evidence to suggest that these trends will not continue, albeit at a relatively moderate pace.

As a result, the operational fighting ability of the PLA has improved dramatically in the past 20 years and will continue to improve with each passing year.  While it is true that the military started from a low base, this should not detract from the PLA's accomplishments.  This improved operational effectiveness has two immediate consequences, one external and one internal.  First, the improving professional fighting ability of the PLA bolsters the credibility of Chinese use of force in the defense of its interests in the region.  This includes its relations with other Asian states as well as its strategic position vis-à-vis the United States.  It is especially relevant to Chinese relations with Taiwan and dynamic of cross-strait interaction.  While no one is suggesting that the Chinese are currently capable of launching a successful invasion of the island, the professionalizing trends of the past 20 years and the acquisition of increasingly sophisticated hardware suggest a rather optimistic trajectory of capabilities vis-à-vis almost any conceivable opponent, including the United States.  In an internal context, the improved professional fighting ability of the PLA strengthens the operational credibility of the military within Chinese policy circles, expanding the strategic options of the political leadership and therefore, by extension, enhancing the position of the military in the national security apparatus.  This domestic credibility, added to the PLA's already vital position in the post-Deng transition, ensures that the voice of the Chinese military will be heard in the foreign policy arena for the foreseeable future, especially regarding issues of coercive diplomacy toward Taiwan and the Spratly Islands.

## IMPLICATIONS FOR SINO-U.S. MILITARY-TO-MILITARY RELATIONS

In the tumult of recent Sino-U.S. relations, one of the most optimistic developments has been the revitalization of military-to-military engagement.  The highlight of this recent round of engagement was the visit of Chinese Defense Minister Chi Haotian to the United States in December 1996 for two weeks of talks and exchanges.  Afterward,

---

[11]See Dennis Blasko, Philip Klapakis, and John F. Corbett, Jr., "Training Tomorrow's PLA: A Mixed Bag of Tricks," *China Quarterly*, No. 146, June 1996, pp. 488–524.

both American and Chinese interlocutors expressed their satisfaction with the visit, which was seen by both sides as a first step toward restoring strategic dialogue between the two nations. One serious obstacle to progress in Sino-U.S. military relations, however, is a continuing lack of understanding about the transformations within the PLA officers corps that are outlined in this report. As military exchanges continue to expand, high-ranking military personnel and civilian policymakers from the U.S. government will come into contact with an increasingly broad cross section of PLA personnel, in particular younger officers with whom U.S. personnel can develop long-term relationships.[12] In order to ensure maximum benefit from these contacts, it is critical that these government officials understand trends in the changing makeup of the highest levels of the Chinese officer corps, since the attitudes and beliefs of this group may significantly affect both China's domestic and external behavior, as well as the military's overall capability.

Should contacts deepen to lower levels of the Chinese military system, U.S. officials and especially U.S. military officers will meet military officers more like themselves: professional, modern, well-educated, and technically capable. Entrance requirements are much higher, as are the standards for promotion. As a result, the PLA officer corps should no longer be viewed in terms of its guerrilla origins, i.e., long on fervor but short on applicable skills. Instead, we should view the senior officer corps in the same way as we viewed the Soviet military, i.e., as a competent military to be respected, although this is not to say that the Chinese military is currently as capable as the Soviet Red Army was at its height or that its intentions are even remotely similar. The reform of the PLA has helped lay the groundwork for a military leadership capable of waging 21st century warfare, even if its equipment still lags well behind advanced global levels. New equipment can be acquired, however, while professional officers capable of maximizing the value of that equipment must be slowly and patiently educated and trained in PME and field envi-

---

[12]One of the main obstacles to improved military-to-military relations is the constant turnover among U.S. officers, which results in very little continuity among high-level personnel in contact with the PLA. As a result, a strong case can be made for encouraging military-to-military contact at the field-grade level, since these ties can develop into useful long-term relationships. This point was suggested to me by Paul Godwin.

ronments. It is these officers who will determine the quality and character of the Chinese military of the future.

## FUTURE RESEARCH

Where do we go from here? Future research should strive to improve the data, both on a micro (individual biographies) and macro (statistical analysis) level. In particular, research in this subfield would be greatly enhanced if we could collect better data on earlier periods, especially the prereform years, so that evolutionary comparisons can be quantified. Within the period in question, more in-depth empirical studies should be undertaken to flesh out the preliminary conclusions in this report. Specifically, it should be possible to match promotions to specific superior/inferior relationships as well as evaluate the source of that relationship. From this lattice, it should be possible to identify genuine networks within the PLA as well as better understand the dynamics of these affiliations.

# Chapter Five

# POSTSCRIPT: THE FIFTEENTH PARTY CONGRESS
# AND THE PLA

The Fifteenth Party Congress promises to be a watershed for the PLA. There are indications that sweeping organizational changes will be announced, centering on the desire of the leadership to significantly reduce the size of the PLA. Reports from Beijing suggest that the PLA will be reduced by an additional 500,000 personnel, primarily through transferring 14 C-class and B-class divisions of the PLA to the People's Armed Police and making deep reductions in head-quarters and educational personnel. Since this reduction will leave many group armies with only 1 viable division, a consolidation of group armies is likely and the distortions in group-army distribution will likely lead to calls for the reduction of the number of military regions, perhaps from seven to four. The result will be a leaner PLA, able to better modernize a smaller force with its current level of fiscal allocation.

The Fifteenth Party Congress will probably also witness major changes in the senior leadership of the PLA. It is expected that CMC Vice-Chairs Liu Huaqing (80) and Zhang Zhen (82) will retire in favor of the new generation of military leadership, embodied by CMC Vice-Chairs Zhang Wannian (68) and Chi Haotian (67). The departure of Liu and Zhang is a watershed in the advance of PLA professionaliza-tion, for it will signal the end of the original revolutionary generation in Chinese military politics. Their replacements generally began their military service near the end of the Anti-Japanese War in 1944, and their formative experience was the Korean War, which was a modern war fought against a technologically advanced opponent. Although Liu was well-known for his advocacy of modernization, the new generations are better equipped to bring the PLA into the 21st century.

As for the personnel decisions among the rank-and-file, the Fifteenth Party Congress may not seem so revolutionary. In fact, the annual winter personnel changes are expected to proceed apace, with retirement-age cadres either promoted to the CMC or retired with full honors. In 1997, these officers include only Beijing MR Commander Li Laizhu (65) and Jinan MR Commander Zhang Taiheng (65).

# BIBLIOGRAPHY

Abrahamsson, Bengt, *Military Professionalization and Political Power*, Beverly Hills, CA: Sage Publications, 1972.

Barany, Zoltan, "Civil-Military Relations in Communist Systems: Western Models Revisited," *Journal of Political and Military Sociology*, Vol. 19, Summer 1991, pp. 75–99.

Baum, Richard, *Burying Mao: Chinese Politics in the Age of Deng Xiaoping*, Princeton: Princeton University Press, 1994.

Blasko, Dennis, Philip Klapakis, and John F. Corbett, Jr., "Training Tomorrow's PLA: A Mixed Bag of Tricks," *China Quarterly*, No. 146, June 1996, pp. 488–524.

Brooks, Timothy, *Quelling the People: The Military Suppression of the Beijing Democracy Movement*, New York: Oxford University Press, 1992.

Bullard, Monte, *China's Military-Political Evolution: The Party and the Military in the PRC, 1960–84*, Boulder, CO: Westview Press, 1984.

————, "People's Republic of China Elite Studies: A Review of the Literature," *Asian Survey*, Vol. 19, No. 8, August 1979, pp. 789–800.

Byrnes, Michael, "The Death of a People's Army," in George Hicks, ed., *The Broken Mirror: China After Tiananmen*, Chicago, IL: St. James Press, 1990, pp. 132–151.

Cheng Hsiao-shih, *Party-Military Relations in the PRC and Taiwan: Paradoxes of Control*, Boulder, CO: Westview Press, 1990.

Colton, Timothy, *Commissars, Commanders, and Civilian Authority: The Structure of Soviet Military Politics*, Cambridge, MA: Harvard University Press, 1979.

*Dangdai Zhongguo jundui de junshi gongzuo [Contemporary Chinese Military Work]*, Beijing: Zhongguo shehui kexue chubanshe, 1989.

Deng Xiaoping, *Selected Works of Deng Xiaoping*, Vol. 1, Beijing: Foreign Language Press, 1982.

*Directory of P.R.C. Military Personalities*, Hong Kong: Defense Liaison Office, U.S. Consulate General, 1989–1994.

Dittmer, Lowell, "Chinese Informal Politics," *The China Journal*, No. 34, July 1995, pp. 1–34.

Dong Lisheng, guest editor, "The Cadre Management System of the Chinese People's Liberation Army (I)," *Chinese Law and Government*, Vol. 28, No. 4, July–August 1995.

Dreyer, June Teufel, "The New Officer Corps: Implications for the Future," *China Quarterly*, No. 46, June 1996, pp. 315–335.

—————, "Regionalism in the People's Liberation Army," in Richard Yang, ed., *CAPS Papers No. 9*, Chinese Council of Advanced Policy Studies, Taipei, Taiwan, May 1995.

Finer, S. E., *The Man on Horseback*, New York: Frederick A. Praeger Press, 1962.

Get, Jer Donald, COL, *What's With the Relationship Between America's Army and China's PLA? An Examination of the Terms of the U.S. Army's Strategic Peacetime Engagement with the People's Liberation Army of the People's Republic of China*, Carlisle Barracks, PA: U.S. Army War College Strategic Studies Institute monograph, 15 September 1996.

Gittings, John, *The Role of the Chinese Army*, Oxford: Oxford University Press, 1967.

Godwin, Paul H. B., *The Chinese Communist Armed Forces*, Maxwell Air Force Base, Alabama: Air University Press, 1988.

—————, "The PLA and Political Control in China's Provinces," *Comparative Politics*, Vol. 9, October 1976, pp. 1–20.

—————, "Professionalism and Politics in the Chinese Armed Forces: A Reconceptualization," in Dale Herspring and Ivan Volyges, eds., *Civil-Military Relations in Communist Systems*, Boulder, CO: Westview Press, 1978, pp. 219–240.

Goerlitz, Walter, *The History of the German General Staff: 1657–1945*, New York: Frederick A. Praeger, 1953.

Henley, Lonnie, "Officer Education in the Chinese PLA," *Problems of Communism*, May–June 1987, pp. 55–71.

Herspring, Dale, and Ivan Volyges, eds., *Civil-Military Relations in Communist Systems*, Boulder, CO: Westview Press, 1978, pp. 219–240.

Huntington, Samuel, *The Soldier and the State: The Theory and Politics of Civil-Military Relations*, New Haven, CT: Yale University Press, 1957.

Janowitz, Morris, *Military Institutions and Coercion in the Developing Nations*, Chicago, IL: University of Chicago Press, 1964.

—————, *The Professional Soldier: A Social and Political Portrait*, Glencoe, IL: Free Press, 1960.

Jencks, Harlan, *From Muskets to Missiles: Professionalism in the Chinese Army, 1945–1981*, Boulder, CO: Westview Press, 1982.

Joffe, Ellis, *The Chinese Army After Mao*, Cambridge, MA: Harvard University Press, 1987.

—————, "Party-Army Relations in China," *China Quarterly*, No. 146, June 1996, pp. 299–314.

—————, *Party and Army: Professionalism and Political Control in the Chinese Officer Corps, 1949–64*, Cambridge, MA: Harvard University Press, 1965.

————, "The PLA and the Economy:  The Effects of Involvement," IIAA/CAPS Conference, "Chinese Economic Reform:  The Impact on Security Policy," Hong Kong, pp. 8–10, July 1994.

Jowitt, Kenneth, "An Organizational Approach to the Study of Political Culture in Marxist-Leninist Systems," *The American Political Science Review*, No. 68, September 1974, pp. 1171–1191.

Lee, Hong Yung, "China's 12th Central Committee:  Rehabilitated Cadres and Technocrats," *Asian Survey*, Vol. 23, June 1983, pp. 673–691.

————, "Mainland China's Future Leaders:  Third Echelon of Cadres," *Issues and Studies*, Vol. 24, No. 6, June 1988, pp. 36–57.

Lee Wei-chin, "Iron and Nail:  Civil-Military Relations in the PRC," *Journal of Northeast Asian Studies*, Vol. 26, January 1991, pp. 132–148.

Li Cheng, *The Rise of Technocracy: Elite Transformation and Ideological Change in Post–Mao China*, Dissertation, Department of Politics, Princeton University, 1992.

Li Cheng and David Bachman, "Localism, Elitism, and Immobilism: Elite Formation and Social Change in Post–Mao China," *World Politics*, Vol. 42, No. 1, October 1989, pp. 64–94.

Li Cheng and Lynn White, "The Army in the Succession to Deng Xiaoping," *Asian Survey*, August 1993, pp. 757–786.

————, "Elite Transformation and Modern Change in Mainland China and Taiwan:  Empirical Data and the Theory of Technocracy," *China Quarterly*, No. 121, March 1990.

————, "The Thirteenth Central Committee of the Chinese Communist Party:  From Mobilizers to Managers," *Asian Survey*, Vol. 28, No. 4, April 1988, pp. 757–786.

Li Nan, "The PLA's Evolving Warfighting Doctrine, Strategy, and Tactics, 1985–95:  A Chinese Perspective," *China Quarterly*, No. 146, June 1996, pp. 443–463.

Liao Gailong and Fan Yuan, eds., *Who's Who in China: Current Leaders*, Vol. 3, Beijing: Foreign Language Press, 1989.

—————, *Who's Who in China: Current Leaders*, Vol. 4, Beijing: Foreign Language Press, 1994.

Lin Chong-Pin, "The Extramilitary Roles of the People's Liberation Army in Modernization: Limits of Professionalization," *Security Studies*, Vol. 1, No. 4, Summer 1992, pp. 659–689.

Manion, Melanie, "The Cadre Management System, Post–Mao: the Appointment, Promotion, Transfer, and Removal of Party and State Leaders," *China Quarterly*, No. 102, June 1985, pp. 203–233.

—————, "Politics and Policy in Post–Mao Cadre Retirement," *China Quarterly*, No. 129, 1992, pp. 1–25.

—————, *Retirement of Revolutionaries in China: Public Policies, Social Norms, Private Interests*, Princeton: Princeton University Press, 1993.

Mao Zedong, "Be Activists in Promoting the Revolution," *Selected Works of Mao Zedong*, Vol. 5, Beijing: Foreign Language Press, 1977.

—————, "On Correcting Mistaken Ideas in the Party," *Selected Works of Mao Zedong*, Vol. 1, Beijing: Foreign Language Press, 1929.

Meyer, Alfred, "Theories of Convergence," in Chalmers Johnson, ed., *Change in Communist Systems*, Stanford, CA: Stanford University Press, 1970, pp. 313–342.

Mills, William deB., "Generational Change in China," *Problems of Communism*, Vol. 32, November–December 1983, pp. 16–35.

Moore, David, and B. Thomas Trout, "Military Advancement: The Visibility Theory of Promotion," *American Political Science Review* 72, 1978, pp. 452–468.

Paltiel, Jeremy, "PLA Allegiance on Parade: Civil-Military Relations in Transition," *China Quarterly*, No. 143, September 1995, pp. 784–800.

Parrish, William, Jr., "Factions in Chinese Military Politics," *China Quarterly*, No. 56, October–December 1973, pp. 667–699.

Peck, B. Mitchell, "Assessing the Career Mobility of U.S. Army Officers: 1950–1974," *Armed Forces and Society*, Vol. 20, No. 2, Winter 1994, pp. 217–237.

Perlmutter, Amos, *The Military and Politics in Modern Times*, New Haven, CT: Yale University Press, 1977.

Perlmutter, Amos, and William LeoGrande, "The Party in Uniform: Toward a Theory of Civil-Military Relations in Communist Political Systems," *American Political Science Review*, Vol. 76, pp. 778–89.

Pye, Lucian, *The Dynamics of Chinese Politics*, Cambridge U.K.: Oelgeschlager, Gunn, and Hain, 1981.

Sandschneider, Eberhard, "Political Succession in the People's Republic of China: Rule by Purge," *Asian Survey*, Vol. 25, No. 6, June 1985, p. 638.

Segal, David, "Selective Promotion in Officer Cohorts," *Sociological Quarterly*, No. 8, 1967, pp. 199–206.

Shambaugh, David, "The Soldier and the State in China: The Political Work System in the People's Liberation Army," *China Quarterly*, No. 127, September 1991, pp. 527–568.

Shichor, Yitzhak, "Demobilization: The Dialectics of PLA Troop Reduction," *China Quarterly*, No. 146, June 1996, pp. 336–359.

Stevens, Gwendolyn, Fred Rosa, Jr., and Sheldon Gardner, "Military Academies as Instruments of Value Change, *Armed Forces and Society*, Vol. 20, No. 3, Spring 1994, pp. 473–484.

Swaine, Michael, "Generations in the PLA," unpublished paper.

——————, *The Military and Political Succession in China: Leadership, Institutions, Beliefs*, Santa Monica, CA: RAND, R-4254-AF, 1992.

——————, *The Role of the Chinese Military in National Security Policymaking*, Santa Monica, CA: RAND, MR-782-OSD, 1996.

Walder, Andrew G., "Career Mobility and the Communist Political Order," *American Sociological Review*, Vol. 60, June 1995, pp. 309–328.

Wang An, *Jundui zheng-guihua jianshe [The Construction of Military "Regularization"]*, Beijing: Guofang daxue chubanshe, 1996.

Weber, Max, *Economy and Society: An Outline of Interpretive Sociology*, Berkeley: University of California Press, 1978.

Wedeman, Andrew, "Bamboo Walls and Brick Ramparts," unpublished dissertation, University of California Los Angeles, 1995.

Whitson, William, and Huang Chen-hsia, *The Chinese High Command: A History of Communist Military Politics, 1927–1971*, London: Praeger Publishers, 1973.

Wilson, Ian, and You Ji, "Leadership by 'Lines': China's Unresolved Succession," *Problems of Communism* 39, January–February 1990, pp. 28–44.

Zang Xiaowei, "The Fourteenth Central Committee of the CCP: Technocracy or Political Technocracy?" *Asian Survey*, August 1993, pp. 787–809.

Zhu Fang, "Party-Army Relations in Maoist China, 1949–76," unpublished dissertation, Columbia University, 1994.

## Review

Pollack, Jonathan, "Review of *The Chinese Army After Mao*, by Ellis Joffe," *American Political Science Review*, Vol. 84, No. 1, March 1990, pp. 339–340.

## Newspaper Articles

"Active Service Regulations Governing Active Duty Officers of the People's Liberation Army," Xinhua Domestic Service, 13 May 1994, in FBIS, 17 May 1994, pp. 35–40.

"Army Colleges To Recruit 10,000 Graduates," Xinhua, 20 June 1995, in FBIS, 21 June 1995, p. 33.

"Be Farsighted in Investing in Trained Persons," *Jiefangjun Bao*, 22 February 1983, p. 1.

Cao Zhi, "CMC Seeks Improvement of Military Cadres," Xinhua Domestic Service, 6 October 1995, in FBIS, 12 October 1995, p. 27.

Chang Hong, "NPC Plans to Reshuffle Top Military Rankings," *China Daily*, 6 May 1994, p. 1, in FBIS, 6 May 1994, pp. 29–30.

Chang Hsiu-fen, "Major Reshuffle of China's Military Hierarchy— Fourth Generation of Military Officers Take Over Important Posts," *Kuang Chiao Ching*, 16 September 1995, No. 276, pp. 16–18, in FBIS, 29 September 1995, pp. 31–32.

Chen Hui, "Three General Departments Promulgate Provisional Regulations Guiding the Enrollment Work of Military Institutes and Academies," *Jiefangjun Bao*, 22 June 1995, p. 1, in FBIS, 21 August 1995, pp. 26–27.

Chen, Kathy, "Jiang Consolidates Control in China With Military Assignments, 5-Year Plan," *The Wall Street Journal*, 29 September 1995, p. 10.

Chen, Mingchi, "Exploring the University Class of the Communist Chinese Junior Commander School," *Zhonggong yanjiu [Studies in Chinese Communism]*, No. 3, 1 March 1985, pp. 74–81, in JPRS-CPS-85-094, 15 September 1985, pp. 87–100.

Cheung, Tai Ming, "Back to the Front: Deng Seeks to De-Politicize the PLA," *FEER*, 29 October 1992, pp. 15–16.

Dai Xingmin and Gai Yumin, "China's Highest Military Institution— Visiting the National Defense University," *Ban Yue Tan*, No. 18, 25 September 1986, pp. 44–47, in FBIS, 15 October 1986, pp. K13–15.

"Do a Good Job on This Important Matter Affecting the Armed Forces as a Whole," *Jiefangjun Bao*, 4 June 1995, p. 1, in FBIS, 2 October 1995, p. 38.

"Education Termed 'Vital' to Modernization," Xinhua, 26 July 1985, in JPRS-CPS-85-085, 22 August 1985, p. 76.

"Effectively Run the Defense University to Train Personnel Well-Versed in Advanced Modern Military Affairs," Xinhua Domestic Service, 18 December 1985, in FBIS, 19 December 1985, pp. K2–K5.

"Elderly Veteran Officers Retire From PLA," Xinhua, 5 March 1985, in FBIS, 6 March 1985, p. K1.

Gai Yumin and Xiong Zhengyan, "National Defense University, China's Highest-Level Military Academy, Founded in Beijing," Beijing Hong Kong Service, 18 December 1985, in FBIS 18 December 1985, p. K1.

Gilley, Bruce, "Air Force Chief's Retirement Goes Unreported," *Eastern Express*, 29 November 1994, p. 8, in FBIS 29 November 1994, pp. 33–34.

—————, "Biggest Shakeup in Years," *Eastern Express*, 22 August 1995, p. 13, in FBIS, 22 August 1995, pp. 35–36.

—————, "'High-Level' Change of Ideology Chiefs Viewed," *Eastern Express*, 5 September 1995, p. 13, in FBIS, 6 September 1995, pp. 33–34.

—————, "Senior PLA Officers Go In Upheaval," *Eastern Express*, 22 August 1995, p. 1, in FBIS, 22 August 1995, pp. 34–35.

Gu Chengwen, "Veteran Cadre Retirement System Improved," *China Daily*, 9 November 1989, p. 1, in FBIS, 9 November 1989, pp. 31–32.

"Half of All Pilots Have College Diplomas," Xinhua, 25 May 1995, in FBIS, 25 May 1995, p. 42.

Hara, Yoshiaki, "PLA Reshuffles Leaders Under Retirement Age System," *Yomiuri Shimbun*, 27 November 1994, p. 5, in FBIS, 28 November 1994, p. 45.

Hollingworth, Clare, "Report on PLA Modernization Efforts," *Pacific Defense Reporter*, 1 September 1985, p. 48.

Hsiao Chung, "Former Commander of Chengdu Military Is Sent to National Defense University for Further Study," *Kuang Chiao Ching*, No. 233, 16 February 1992, pp. 14–17, in FBIS, 3 March 1992, pp. 34–35.

Hu Wan, "The Cradle of Artillery Technical Talent—A Sketch of the PLA Artillery Technical College," *Anhui Huabao [Anhui Pictorial]*, No. 4, 1 August 1985, p. 6, in JPRS-CPS-85-114, 15 November 1985, pp. 131–132.

Isokawa, Tomoyoshi, "PLA Reportedly Carries Out Personnel Changes," *Asahi Shimbun*, 10 August 1995, p. 9, in FBIS, 11 August 1995, p. 14.

Lam, Willy Wo-Lap, "Chengdu, Jinan Military Region Leaders Retire," *South China Morning Post*, 17 December 1994, p. 10, in FBIS, 19 December 1994, pp. 31–32.

———, "Further Changes Planned," *South China Morning Post*, 7 October 1995, p. 8, in FBIS, 10 October 1995, pp. 46–47.

———, "PAP Undertakes Thorough 'Changing of the Guards,'" *South China Morning Post*, 24 July 1996.

———, "Reasons for Military Reshuffling Examined," *South China Morning Post*, 5 September 1995, p. 1, in FBIS, 6 September 1995, pp. 32–33.

———, "Senior Generals Involved in Foreign Affairs," *South China Morning Post*, 25 June 1994, p. 10, in FBIS, 27 June 1994, pp. 40–41.

———, "Top PLA Posts Go To High-Tech Experts," *South China Morning Post*, 4 December 1996, p. 1.

"Lanzhou Air Force Leaders Set Example in Reorganization," *Lanzhou Gansu Provincial Service*, 13 June 1985, in JPRS-CPS-85-071, 22 July 1985, p. 145.

Liu Huinian and Zhang Chunting, "To Run the Army Well, It Is First Necessary to Run the Military Colleges Well—Xiao Ke on Building of Military Colleges," *Liaowang*, No. 7, 20 July 1983, in JPRS-CPS-84-273, No. 454, 8 September 1983, p. 74.

Liu Ping-chun, "Jiang Zemin Orders Retirement of Five Navy, Air Force Generals," *Ming Pao*, 3 December 1996, p. A12.

Liu Yuming, "On Persistently Exploring People's War on High-Tech Terms," *Guofang [National Defense]*, 15 October 1994, No. 10, p. 8, in FBIS, 18 September 1995, pp. 19–20.

Lo Ping, "CPC Military Attacks Ministry of Foreign Affairs," *Zhengming*, No. 201, 1 July 1994, pp. 6–8, in FBIS, 26 July 1994, pp. 33–36.

Lo Ping and Li Tzu-ching, "The Whole Story on Promoting Military Officers to Generalship," *Zhengming*, No. 201, 1 July 1994, pp. 11,12, in FBIS, 27 July 1994, pp. 38–39.

Luo Tongsong, Wang Jin, and Gai Yumin, "Leaders Attend Founding of Defense University," Xinhua, 15 January 1986, in FBIS, 17 January 1986, pp. K5–7.

Ma Tien-long, "Series of Unusual Changes Take Place in Military Systems in Both Taiwan and the Mainland," *Pingguo Ribao*, 2 September 1995, p. A14, in FBIS, 8 September 1995, pp. 32–33.

Massonnet, Phillippe, "Chinese Army Names 19 New Generals," *Agence France Presse English Wire*, 9 June 1994, in *China News Digest*, 11 June 1994.

"Military Commission Promotes 19 Army Officers to General," Xinhua, 8 June 1994, pp. 31–33, in FBIS, 9 June 1994, pp. 31–32.

"Military Urged to Support Academy Reform," Xinhua, 2 June 1986, in FBIS, 3 June 1986, pp. K25–26.

"PLA Attracts Graduates With Advanced Degrees," Xinhua, 22 July 1994, in FBIS, 22 July 1994, p. 20.

"PLA Political Unit Makes Decision on Education," Xinhua, 4 May 1983, in FBIS 10 May 1983, pp. K8–9.

"PLA Reportedly Carries Out Personnel Changes," *Asahi Shimbun*, 10 August 1995, p. 9, in FBIS, 11 August 1995, p. 14.

"Regulations for Military Service of Active Duty Officers of the Chinese People's Army," Xinhua, in FBIS, 8 September 1988, pp. 37–41.

Segal, David, "Selective Promotion in Officer Cohorts," *Sociological Quarterly* 8, 1967, pp. 199–206.

"Senior PLA Officers Retire From Leading Posts," Xinhua, 29 December 1984, p. K2.

Shang Dajia, "Introduction to China's Female Generals," *Beijing Review*, No. 32, 8–14 August 1994, pp. 11–19.

Sun Maoqing, "Air Force Implements Retirement Plan for Pilots," Xinhua Domestic Service, 1 December 1985, in FBIS, 2 December 1985, pp. K24–25.

Sun Maoqing, "Air Force Sees Age, Education Changes in Leaders," Xinhua Domestic Service, 12 September 1994, in FBIS, 20 September 1994, pp. 23–24.

Taft, Sheila, "China Snarls at the World As Military Calls More Shots," *Christian Science Monitor*, 22 August 1995, p. 1.

Tai Ming Cheung, "Back to the Front:  Deng Seeks to De-Politicize the PLA," *FEER*, 29 October 1992, pp. 15–16.

Tseng Hui-yen, "China's Military Power Gradually Falls Into Hands of 'Shandong Faction':  Officers Seek Promotion by Advocating Military Expansion Strategy," *Lien Ho Pao*, 21 October 1994, p. 10, in FBIS, 21 October 1994, pp. 34–35.

Wang An and Yang Minqing, "PLA Modernizes Military Education System," Xinhua Domestic Service, 14 September 1986, in FBIS, 17 September 1986, p. K1.

Wang Xian, "Analysis of the Chinese Communists' Establishment of a 'National Defense University,'" *Studies of Chinese Communism*, Vol. 20, No. 8, 15 August 1986, pp. 83–90, translated in JPRS-CPS-86-078, 9 October 1986, pp. 1–15.

Wei Houmin and Fu Yongguo, "Zhang Zhen Inspects Military Academies," *Beijing Central People's Radio*, "News and Press Review," 16 September 1995, in FBIS, 26 September 1995, p. 26.

WuDunn, Sheryl, "War Astonishes Chinese and Stuns Their Military," *New York Times*, 20 March 1991, p. A13.

Xiao Ke, "Guidelines for Building Colleges and Academies of Our Army," *Renmin Ribao*, 3 October 1983, in FBIS, 5 October 1983, pp. K7–10.

Xu Jingyue and Jing Shuzhan, "PLA Adopts System for Promoting High-Ranking Cadres," Xinhua Domestic Service, 24 June 1994, in FBIS, 28 June 1994, pp. 41–42.

Xu Shen, "Air Force Pilots Are Now Full-Time Undergraduate Students," *Jiefangjun Bao*, 15 October 1994, p. 1, in FBIS, 27 October 1994, p. 48.

Yueh Shan, "Fifty-Eight Generals to Retire," *Cheng Ming*, No. 207, 1 January 1995, pp. 18–19, in FBIS, 16 February 1995, pp. 42–43.

Zhang Zenan, "Rear Admiral Li Dingwen, Director of the Naval Command Academy, Speaks of the Chinese Navy's Highest Institution of Higher Learning," *Jianchuan Zhihshi [Naval and Merchant Shipping]*, No. 8, 8 August 1992, pp. 2–3, in JPRS-CAR-93-001, 8 January 1993, pp. 51–54.

Zhao Su, "Qin Jiwei Praises, Encourages, Retired PLA Cadres," Xinhua Domestic Service, 12 February 1985, in JPRS-CPS-85-020, 4 March 1985.

Zhu Ling, "China's Army is Gearing Itself for Modern Warfare," *China Daily*, 11 June 1983, in FBIS, 13 June 1983, pp. K29–30.